FUNDAMENTALS OF MICROOPTICS
Distributed-Index, Microlens, and Stacked Planar Optics

FUNDAMENTALS OF MICROOPTICS

Distributed-Index, Microlens,
and Stacked Planar Optics

K. IGA
Y. KOKUBUN
M. OIKAWA

Tokyo Institute of Technology
Nagatsuta, Midoriku, Yokohama, Japan

1984

ACADEMIC PRESS, INC.

(Harcourt Brace Jovanovich, Publishers)

Tokyo   Orlando   San Diego   New York
London   Toronto   Montreal   Sydney

OHM

Tokyo・Osaka・Kyoto

ENGLISH LANGUAGE EDITION COPYRIGHT © 1984
BY OHMSHA, LTD.

Copublished by
OHMSHA, LTD.
1-3 Kanda Nishiki-cho, Chiyoda-ku, Tokyo 101
and

ACADEMIC PRESS JAPAN, INC.
Hokoku Bldg. 3-11-13, Iidabashi, Chiyoda-ku, Tokyo 102

United States Edition published by ACADEMIC PRESS, INC.
Orlando, Florida 32887

United Kingdom Edition published by ACADEMIC PRESS, INC. (LONDON) LTD.
24/28 Oval Road, London NW1 7DX

Distributed in Japan by Ohmsha, Ltd., and Academic Press Japan, Inc.
Distributed outside Japan by Academic Press, Inc.

Library of Congress Cataloging in Publication Data

Iga, Ken'ichi, Date
    Fundamentals of microoptics.

    Includes index.
    1. Integrated optics. 2. Optical wave guides.
I. Kokubun, Y. II. Oikawa, M. III. Title.
TA1660.I35 1984   621.36'9   83-12295
ISBN 0-12-370360-3

PRINTED IN THE UNITED STATES OF AMERICA

84 85 86 87   9 8 7 6 5 4 3 2 1

To Professor Yasuharu Suematsu

Contents

Preface

Lightwave electronics, including optical fiber communications and optoelectronic devices, has become one of the most promising fields of engineering since the laser first appeared. Although some light source devices such as semiconductor lasers and optical components have reached practical levels, it must be noted that we are still far from utilizing high-frequency lightwaves to their fullest advantage. In present systems, optical components are composed of discrete devices; this corresponds to the early stages of electronics before integrated circuits.

When we look at the very rapid progress occurring in electronics, we recognize that electron device integration technology and large scale integration have been contributing a great deal. In the field of lightwave electronics, on the other hand, we must develop the ability to compose optical components without optical axis alignment or hand-made fabrication if we want to proceed to the industrial level with high performance and reliability, mass production, and low cost.

When we consider the history of electronic devices, it would seem that the only way to do this is to develop the technology to integrate and monolithically fabricate optical components. Unfortunately, no one knows the best way to this final goal; small improvements alone will not be enough and fundamental innovations are needed.

We know, however, that microoptic components composed of microlenses give us excellent performances, i.e., low insertion loss, very small reflection, high extinction ratio, and so on. Problems existing in microoptic components include: (i) alignment of optical axis, (ii) complexity of surface treatment, (iii) fabrication of discrete devices, and (iv) coupling of optical fibers. Therefore, it is necessary to introduce some innovation to utilize the superiority inherent in microoptics.

This book presents one possible scheme, namely, two-dimensionally arrayed components based on a planar microlens array. Through the discovery of this new configuration not only the basic theoretical concept but also detailed microoptic technologies have achieved a lasting technical position.

In this book we begin with a presentation of the theoretical methods related to microoptics, especially distributed-index optics, fabricational technologies, and measurement techniques, and finally propose a stacked planar optics, that is, the two-dimensional array optics concept.

Scientists and engineers in the field of lightwave communications and optoelectronics will find this book of interest. Those concerned with optics may also be interested in this new field. Students in electrical and electronics engineering and applied physics can use this volume as a textbook or a reference book.

We hope this book will serve to give the basic ideas and theoretical background not only in the area of lightwave communications but also for expanding technical fields of optoelectronics, such as optical disks, imaging systems, and lightwave sensing.

Acknowledgments

We would like to sincerely acknowledge Professor Yasuharu Suematsu of Tokyo Institute of Technology for his continuing encouragement to our study and research. We would like to thank Professor Toshiharu Tako (presently with Science University of Tokyo) and Professor Junpei Tsujiuchi of Tokyo Institute of Technology for their helpful advice. We also thank Drs. Ichiro Kitano, Ken Koizumi, Tetsuya Yamazaki, and Kouichi Nishizawa of Nippon Sheet Glass Company for useful discussions. We thank Noboru Yamamoto (presently with Canon), Shigeru Ohshima (presently with Toshiba), Shigeyoshi Misawa, Junichi Banno (presently with IBM), and other research students for their assistance in microoptics research. Many thanks are due Mrs. Hiroko Watanabe and Miss Hiromi Kurokawa who typed most of the manuscript and the staff of Ohmsha, Ltd., who gave us editorial help.

CHAPTER 1

Introduction

1.1 OPENING REMARKS

In this chapter we introduce the meaning of microoptics and how it is applied to engineering. To deal with lightwaves in lightwave systems, several kinds of optical circuits and components have been used, including micro-optics, guided-wave optics, and integrated optics. Conventional microoptics consist of discrete microlenses and other optical devices. Guided-wave optical components basically utilize dielectric planar waveguides or optical-fiber circuitries. Integrated optics are used to construct optical circuits by a mono-lithic fabrication process. We shall briefly review these optical devices before detailing the theory and technology of microoptics.

1.2 MICROOPTICS

As mentioned in the Preface, a number of optical devices used in optical-fiber communications and optoelectronics systems have progressed beyond prototype and reached the level of production. In the communication and electronics fields, there are optical components that consist of very small lenses for focusing, imaging, branching, and transmitting lightwaves. This classification of optical components, or optics, is known as microoptics,[1] and it is becoming increasingly important.

For focusing and imaging components, microlenses such as distributed-index (DI) or gradient-index (GRIN) rod lenses,[2,3] tiny spherical lenses,[4] Fresnel lenses,[5] and similar optics have been utilized. Among these, the distributed-index lens has an advantage in that other optical components can be connected with the surface of the lens, since the focusing effect originates not from the shape of the lens surface but from the internal index distribution, as shown in Fig. 1.2-1. Another merit of the DI lens is the potential for forming a real-imaging system with a magnification of $+1$ using only one singlet lens.[6]

Such microlenses with 0.1- to 1-mm diameters have been fabricated using ion exchange,[7] electromigration,[8] chemical vapor deposition,[9] diffusion polymerization,[10,11] and various other techniques introduced for fabricating optical fibers and dielectric waveguides. Also, etching and lithographic

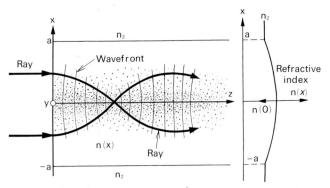

FIGURE 1.2-1.   Ray focusing by a distributed-index medium.

technology have only recently been transferred[8,9] into optoelectronic devices from the electronics field. Various materials are being tested for forming optical components: glass, silica, plastics, semiconductors, electro- and magnetooptic crystals (according to the application).

There are several lightwave components used in systems today.[12] In lightwave communications, couplers are utilized to combine light sources and fibers, branching circuits, wavelength multiplexing and demultiplexing circuits, and so on. Components are divided into those for multimode fibers and those for single-mode fibers.

In optical memory and laser disk systems,[13] microlenses focus laser light into a spot nearly 1 μm in diameter, allowing the reading or memorizing of pulse signals from/to optical disks. A lightweight lens is needed because motional feedback is essential to compensate for any irregular movement in the disks. In copy machines,[14] lens arrays are introduced to make them very compact, and the DI lens array forms images of the original with one magnification. Some facsimiles[15] utilize the same principle for transforming images; DI imaging rods are used in some diagnostic tools in the medical field.[16] In the near future, microoptics might find new applications in cameras, printers, and various optoelectronics systems.

1.3 GUIDED-WAVE OPTICS

The microoptics introduced in the preceding section utilize a discrete lens system for the purpose of focusing, transmitting, and imaging. Guided-wave optics, on the other hand, use a dielectric waveguide to deal with lightwaves. There are two configurations of guided-wave optics—one uses optical-fiber circuitries and the other, planar waveguides.

In the latter case, the light guiding property of thin films on substrates is used, sometimes with narrow strips which enable lightwaves to be confined in the transverse direction, as shown in Fig. 1.3-1. One main objective of guided-wave optical technology is to construct so-called integrated optics,[17] but it is very difficult for the optical circuit to be integrated onto one chip. However, it is a very effective method of utilizing the guided-wave configuration for forming high-performance lightwave components. Various materials are utilized in the planar waveguide, for example, optical glass, plastics, semiconductors, and electrooptic, magnetooptic, and liquid crystals. These materials have the required characteristics (e.g., changeable refractive index and anisotropy) and are used to create functional devices, such as optical modulators, switches, signal processors, phase conjugates, and nonlinear optical devices. Their waveguide configuration is useful for confining light in a narrow space, which gives high field intensity. Guided-wave optics will find wider application when integrated in lightwave circuits and the problem of coupling them to optical fibers and lasers can be solved.

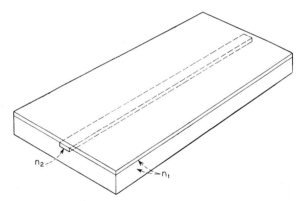

FIGURE 1.3-1.   Waveguide optics where n_2 and n_1 denote the refractive index of the core and cladding, respectively. (After Miller.[17])

Optical fiber circuitry[18] consists of manufactured optical fibers, taking advantage of their characteristics to transmit (i) amplitude information, (ii) phase information, (iii) spatial information, and (iv) light-wave energy without being subjected to environmental disturbances. Optical sensors using optical fibers are rapidly extending their applications to the monitoring of acoustic waves, pressure, magnetic fields, electric fields, acceleration, rotational speed, and many other physical phenomena.

1.4 INTEGRATED OPTICS

As the utilization of lightwave systems has become more widespread, the need to integrate devices with each other and to drastically improve the performance of components has become well recognized. The term "integrated optics" was first suggested by Miller in 1969.[17] At that time, some persons wanted to regard optical semiconductor devices as being those composed of dielectric waveguides, or they tried to construct them using guided-wave structures. Miller's proposal to utilize planar waveguides gathered much attention from lightwave engineers who were engaged in communications and optoelectronic devices. Some of them succeeded in producing very compact devices with the help of the guided-wave concept. However, there have been only a few "integrated" lightwave components, although many researchers have devoted a lot of effort to the integration of optical devices. Therefore, devices that utilize a guided-wave configuration may be regarded as "guided-wave optics" for the previously mentioned reason. Moreover, there are few guided-wave optical device currently in practical use.

In 1979 Yariv and his co-workers demonstrated the integration of lightwave and electron devices such as lasers, detectors, and transistors, as shown in Fig. 1.4-1. They proposed the monolithic formation of optical repeaters, transmitters, receivers, and similar complex systems. (Now, the idea of integrating semiconductor devices seems to be a physical necessity.) Kaminow and co-workers,[20] on the other hand, first tried the use of gratings on a waveguide to achieve lasing inside the waveguide grating, which was proposed by Miller. This trial was, unfortunately, unsuccessful and only a "dark laser" remained in their hands. Soon after this, Kogelnik and Shank[21] succeeded in the oscillation of the first "distributed feedback" (DFB) dye laser, which they named as such. Yariv and Nakamura[22] introduced semiconductor material for constructing DFB lasers. Thus the use of waveguide gratings opened up a new method for creating integrated lasers.

Another important technique may be the use of semiconductor waveguides placed next to or connected to a laser device. There have been several such waveguide devices or methods, such as the butt joint,[23] taper coupler,[24] evanescent coupler,[25] and twin-guide or directional coupler.[26] The integration of many optical and electron devices that have thus far been connected by built-in waveguides may find further applications in the future. Semiconductor integrated optics is one of the most promising concepts in the field of integrated optics. To summarize, some areas of application of semiconductor integrated optics are

(1) integration of lightwave devices into one chip by including mode-controlling or frequency-controlling devices to improve performance,
(2) integration of lightwave and electron devices,
(3) array formation of lightwave devices,
(4) integration of switching or exchanging components.

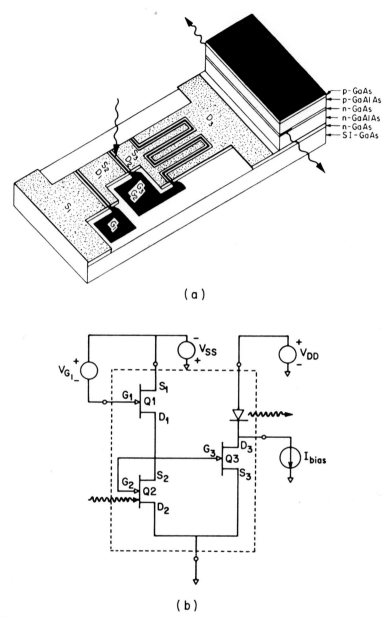

(a)

(b)

FIGURE 1.4-1.   Integration of electronic and optical devices. The basic configuration of an
optical receiver is realized in this device. The detector (DET) receives the optical signal transmitted
and the electronic circuits reshapes the pulses. The laser diode (LD) is driven by the reshaped
current signal and the output optical pulse is transmitted again. (After Yust, Bar-Chaim,
Izadpanah, Margalit, Ury, Wilt, and Yariv.[19])

1.5 COMPARISON OF LIGHTWAVE CIRCUITS

In the preceding sections some characteristics of microoptics represented by discrete devices and guided-wave optics based on the planar dielectric waveguide were discussed. In this section the problems of the associated circuitry will be discussed. In Table 1.5-I, the problems and difficulties involved with discrete optics and guided-wave optics are summarized, although they might be biased by the authors' personal opinion. In any case, there are many problems with both configurations and they need to be overcome by introducing some other concepts or by improving current technology.

One important objective of writing this book is to present "stacked planar optics" as a solution for these problems. This concept is based on the utilization of the 2-D microlens array shown in Fig. 1.5-1. The authors have reported this concept in previously published papers,[27-29] but the details are reproduced in the last chapter of this book. It is hoped that this form of optics will be helpful in opening up new lightwave technology.

TABLE 1.5-I

ADVANTAGES OF STACKED PLANAR OPTICS

	Optical system		
Classification of encountered problem	Discrete	Waveguide	Stacked
Fabrication process	Yes	No	No
Surface preparation	Yes	Yes	No
Optical alignment	Yes	Yes	No
Coupling	No	Yes	No
Integration of different materials	No	Yes	No
Single–multi-compatibility	No	Yes	No
Polarization preference	No	Yes	No
Mass-production	Yes	No	No
2-D array	Yes	Yes	No
large-scale optics ←			

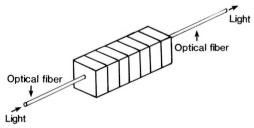

FIGURE 1.5-1. Stacked optical circuit (discrete component). (After Iga, Oikawa, Misawa, Banno, and Kokubun.[29])

REFERENCES

1. T. Uchida and I. Kitano, *Electron. Eng.* (Tokyo) **27**, 22 (1969).
2. T. Uchida, M. Furukawa, I. Kitano, K. Koizumi, and H. Matsumura, *IEEE J. Quant. Electron.* **QE-6**, 606 (1970).
3. A. D. Pearson, W. G. French, and E. G. Rawson, *Appl. Phys. Lett.* **15**, 76 (1979).
4. M. Saruwatari *et al.*, *Electron. Lett.* **116**, 45 (1980).
5. T. Suhara, K. Kobayashi, H. Nishihara, and J. Koyama, *Appl. Opt.* **21**, 1966 (1982).
6. K. Matsushita and M. Toyama, *Appl. Opt.* **19**, 1070 (1980).
7. I. Kitano, K. Koizumi, H. Matsumura, T. Uchida, and M. Furukawa, *Jpn. J. Appl. Phys.* **39**, 63 (1970).
8. M. Oikawa, K. Iga and T. Sanada, *Electron. Lett.* **17**, 452 (1981).
9. G. D. Khoe, H. G. Kock, J. A. Luijendik, C. H. J. van den Brekel, and D. Kueppers, Digest of 7th ECOC. No. 7.6, Copenhagen (1981).
10. Y. Ohtsuka, *Appl. Phys. Lett.* **23**, 247 (1973).
11. K. Iga, K. Yokomori, and T. Sakayori, *Appl. Phys. Lett.* **26**, 578 (1975).
12. T. Uchida and K. Kobayashi, *Optical Devices and Fibers*. Ohmsha, Tokyo and North Holland, Amsterdam, (1982).
13. T. Miyazawa, K. Okada, T. Kubo, I. Kitano, K. Nishizawa, and K. Iga, *Appl. Opt.* **19**, 1113 (1980).
14. M. Kawazu and Y. Ogura, *Appl. Opt.* **19**, 1070 (1980).
15. K. Komiya, M. Kanzaki, Y. Hatate, and T. Yamashita, *Trans. IECE Jpn.* **IE80**, 72 (1980).
16. K. Nishizawa, *Appl. Opt.* **19**, 1052 (1980).
17. S. E. Miller, *Bell Syst. Tech. J.* **48**, 2059 (1969).
18. J. J. Pan, Digest of Conference on Lasers and Electro-Optic Systems, San Diego, No. ThE 3 (1976).
19. M. Yust, N. Bar-Chaim, S. H. Izadpanah, S. Margalit, I. Ury, D. Wilt, and A. Yariv, *Appl. Phys. Lett.* **35**, 795 (1979).
20. I. P. Kaminow, pers. comm.; I. P. Kaminow, H. P. Weber, and E. A. Chandross, *Appl. Phys. Lett.* **18**, 497 (1971).
21. H. Kogelnik and C. V. Shank, *Appl. Phys. Lett.* **18**, 152 (1971).
22. M. Nakamura, A. Yariv, H. W. Yen, S. Somekh, and H. L. Garvin, *Appl. Phys. Lett.* **22**, 515 (1973).
23. C. E. Hurwitz, J. A. Rossi, J. J. Hsieh, and C. M. Wolfe, *Appl. Phys. Lett.* **27**, 241 (1975).
24. J. L. Merz, R. A. Logan, W. Wiegman, and A. C. Gossard, *Appl. Phys. Lett.* **26**, 337 (1975).
25. W. T. Tsang and S. Wang, *Appl. Phys. Lett.* **28**, 596 (1976).
26. Y. Suematsu, M. Yamada, and K. Hayashi, *Proc. IEEE*, **63**, 208 (1975).
27. K. Iga, M. Oikawa, and J. Banno, *Top. Meet. Integ. Guide-Wave Opt.* **FB6-1** (1982).
28. M. Oikawa, K. Iga, and S. Misawa, *Electron. Lett.* **18**, 316 (1982).
29. K. Iga, M. Oikawa, S. Misawa, J. Banno, and Y. Kokubun, *Appl. Opt.* **21**, 3456 (1982).

CHAPTER 2

Ray Theory

2.1 OPENING REMARKS

This chapter examines the transmission of a light ray through a distributed-index medium. There have been a number of textbooks describing ray optics[1-3] of lenses and DI media,[4] but we feel it is worthwhile to prepare a self-consistent ray theory for possible application to lightwave components which are to be treated later in this book.

2.2 DISTRIBUTED-INDEX LENSES AND EXPRESSION OF INDEX DISTRIBUTION

As will be discussed in the following chapters, optical components with a distributed index may play an important role in the microoptics area. There have been various ways of expressing the index distribution of such devices. One method is a power series expansion of the refractive index with respect to coordinates.[5] Another is also a power series expansion, but of the squared refractive index with respect to coordinates.[6] The advantage of the latter method is described in Ref. 2, and this method will be used in this book as described in this section.

The expression of the refractive-index distribution in the distributed-index or gradient-index medium that we first presented is in the form

$$n^2(r) = n^2(0)[1 - (gr)^2 + (hr)^4] \qquad (2.2\text{-}1)$$

for a circularly symmetric fiber or rod, where r denotes the transverse distance and g and h are constants that express the index gradient.[7] A dielectric constant is used rather than a refractive index, since the wave equation includes a dielectric constant or squared represent expression,[8] (after Ref. 9). That is,

$$n^2(r) = n^2(0)[1 - (gr)^2 + h_4(gr)^4 + h_6(gr)^6 + \cdots], \qquad (2.2\text{-}2)$$

where g is a focusing constant expressing gradient index and h_4 and h_6

represent higher-order terms of index distribution and are closely related to aberration. In Eqs. (2.2-1) and (2.2-2), $n(0)$ expresses the index at the center axis when $r = 0$. Advantages of this expression can be summarized as follows:

(1)   We can substitute Eq. (2.2-2) into the wave equation, which contains an inhomogeneous dielectric constant

$$\nabla^2 E + k_0^2 n^2(r)E + \{\text{grad } n^2(r) \text{ terms}\} = 0, \tag{2.2-3}$$

where $k_0 = 2\pi/\lambda$. This type of wave equation, which can usually be reduced to a scalar form, is commonly applied to an optical fiber that has a distributed index at the core.

(2)   Equation (2.2-1) can also be utilized to ascertain the ray trajectory $r(x, y)$ by solving the ray equations

$$\frac{d^2 x}{dz^2} = \frac{1}{2n^2(r_i) \cos^2 \gamma_i} \frac{\partial n^2(x, y)}{\partial x}, \tag{2.2-4a}$$

$$\frac{d^2 y}{dz^2} = \frac{1}{2n^2(r_i) \cos^2 \gamma_i} \frac{\partial n^2(x, y)}{\partial y}, \tag{2.2-4b}$$

where z denotes the axial distance, $r_i (= \sqrt{x_i^2 + y_i^2})$ the incident ray position, and γ_i the direction cosine of the incident ray.[10–12] A solution to the meridional ray has been obtained in the closed form.[10]

(3)   It is convenient for contrasting with the α-class expression,[13] which is common in optical multimode fibers, that is

$$n^2(r) = \begin{cases} n^2(0)[1 - 2\Delta(r/a)^\alpha] & \text{if} \quad r \le a, \\ n_2^2 & \text{if} \quad r > a, \end{cases} \tag{2.2-5}$$

where $\Delta = [n(0) - n_2]/n(0)$.

(4)   It is convenient for relating to the density of particles contributing to the refractive index, that is, the Clausius–Mossoti relation or the Lorenz–Lorentz relation including n^2 terms rather than n terms.[14]

Recently we proposed and demonstrated the feasibility of a planar microlens having 3-D index distribution.[15] To solve the ray equation in the 3-D problem, the form of Eq. (2.2-1) is extended in a matrix form as

$$n^2(r, z) = n^2(0, 0)[1, gz, (gz)^2, (gz)^3, \ldots](N) \begin{bmatrix} 1 \\ (gr)^2 \\ (gr)^4 \\ \vdots \end{bmatrix}, \tag{2.2-6}$$

where the index matrix is given by

$$
N = \begin{bmatrix}
1 & -1 & v_{04} & v_{06} & \cdots \\
v_{10} & v_{12} & v_{14} & v_{16} & \cdots \\
v_{20} & v_{22} & v_{24} & v_{26} & \cdots \\
v_{30} & v_{32} & v_{34} & v_{36} & \cdots \\
\vdots & \vdots & \vdots & \vdots & \ddots
\end{bmatrix}.
\tag{2.2-7}
$$

To match this to Eq. (2.2-2) for an axially symmetric case, we can put $v_{04} = h_4$, $v_{06} = h_6$, and so on.

In a special case where the index matrix is of the form

$$
N = \begin{bmatrix}
1 & -1 & v_{04} & v_{06} & \cdots \\
v_{10} & & & \\
v_{20} & & 0 & \\
\vdots & & &
\end{bmatrix},
\tag{2.2-8}
$$

the ray equation reduces to a simple integral expression.[16] Thus we have

$$
n^2(r, z) = n^2(0, 0)[R(r) + Z(z)].
\tag{2.2-9}
$$

2.3 RAY EQUATION

A. BASIC RAY EQUATION

Light propagation is simply described by a ray which indicates a path of light energy. Actually, sometimes a very thin beam from a laser or collimated light through a pinhole appears as a "ray" that is certainly indicating the path of light energy. In a conventional optical system, which is constructed with lenses, mirrors, prisms, and so on, a ray is represented by a straight line and being refracted or reflected at a surface where refractive index changes. On the other hand, in a distributed index medium, a ray does not follow a straight line; rather it takes curved trajectory as if it is affected by a force toward the higher refractive index. The ray trajectory in a distributed index medium is calculated from ray equations which are second-order partial differential in nature. Here, the ray is mathematically defined as a curve perpendicular to the wavefront being calculated from a limited case in that $\lambda \to 0$.[1] According to this definition, the tangential direction of the ray corresponds to the direction of the time-averaged Poynting vector. This definition is consistent with Fermat's principle[1,2] i.e., the ray trajectory takes the shortest optical path length. We derive ray equations from Fermat's principle, since the variational method can be used from which we can derive ray equations easily in any desired coordinate system. The optical pass length is defined by

$$\mathscr{L} = \int_{P_1}^{P_2} n \, ds, \tag{2.3-1}$$

where n is the refractive index of the medium of interest (see Fig. 2.3-1). Using coordinate variables $x(\tau)$, $y(\tau)$, and $z(\tau)$ as functions of independent parameter τ, we have the equation of optical pass length

$$\mathscr{L} = \int_{\tau_1}^{\tau_2} n(x, y, z)\sqrt{x'^2 + y'^2 + z'^2} \, d\tau. \tag{2.3-2}$$

Here, ds in Eq. (2.3-1) is rewritten as $\sqrt{x'^2 + y'^2 + z'^2} \, d\tau$, and the prime means differentiation with respect to τ, where τ_1 and τ_2 are, respectively, the initial and final values of τ. To find the ray trajectory, we use Euler's equation, i.e., when the functional F is a function of x, y, z, x', y', z', and τ:

$$F = F(x, y, z, x', y', z'; \tau) \tag{2.3-3}$$

and integration of F takes stationary value, relations among x, y, and z are described by the following three equations:

$$\frac{d}{d\tau}\left(\frac{\partial F}{\partial x'}\right) = \frac{\partial F}{\partial x}, \tag{2.3-4a}$$

$$\frac{d}{d\tau}\left(\frac{\partial F}{\partial y'}\right) = \frac{\partial F}{\partial y}, \tag{2.3-4b}$$

$$\frac{d}{d\tau}\left(\frac{\partial F}{\partial z'}\right) = \frac{\partial F}{\partial z}, \tag{2.3-4c}$$

Since the functional of Eq. (2.3-2) is

$$F(x, y, z, x', y', z'; \tau) = n(x, y, z)\sqrt{x'^2 + y'^2 + z'^2}. \tag{2.3-5}$$

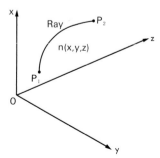

FIGURE 2.3-1. Ray propagation in a DI medium.

Euler's equations satisfying Eq. (2.3-4) are written as (with $n_x = \partial n/\partial x$)

$$\frac{d}{d\tau}\frac{nx'}{\sqrt{x'^2 + y'^2 + z'^2}} = n_x\sqrt{x'^2 + y'^2 + z'^2}, \qquad (2.3\text{-}6a)$$

$$\frac{d}{d\tau}\frac{ny'}{\sqrt{x'^2 + y'^2 + z'^2}} = n_y\sqrt{x'^2 + y'^2 + z'^2}, \qquad (2.3\text{-}6b)$$

$$\frac{d}{d\tau}\frac{nz'}{\sqrt{x'^2 + y'^2 + z'^2}} = n_z\sqrt{x'^2 + y'^2 + z'^2}. \qquad (2.3\text{-}6c)$$

If we give certain meaning to the parameter $d\tau = ds/n$, we obtain differential ray equations in the Cartesian coordinate:

$$\frac{d^2x}{d\tau^2} = \frac{\partial}{\partial x}\frac{1}{2}n^2, \qquad (2.3\text{-}7a)$$

$$\frac{d^2y}{d\tau^2} = \frac{\partial}{\partial y}\frac{1}{2}n^2, \qquad (2.3\text{-}7b)$$

$$\frac{d^2z}{d\tau^2} = \frac{\partial}{\partial z}\frac{1}{2}n^2. \qquad (2.3\text{-}7c)$$

When we treat a DI rod lens with the index expressed by Eq. (2.2-2), the partial derivative of n^2 with respect to z is zero when we choose the z axis as the optical axis. Equation (2.3-7c) is integrated and we have

$$dz/d\tau = C_i. \qquad (2.3\text{-}8)$$

For the ray incident at $x = x_i$, $y = y_i$, and $z = 0$, we have

$$C_i = n(x_i, y_i, 0)\cos\gamma_i. \qquad (2.3\text{-}9)$$

Since $d\tau = ds/n$, here $\cos\gamma_i$ is a directional cosine of the ray at $z = 0$ with respect to the z axis.

Elimination of τ from Eq. (2.3-7) using Eqs. (2.3-8) and (2.3-9) gives

$$\frac{d^2x}{dz^2} = \frac{1}{2n^2(r_i)\cos^2\gamma_i}\cdot\frac{\partial n^2(x, y)}{\partial x}, \qquad (2.3\text{-}10a)$$

$$\frac{d^2y}{dz^2} = \frac{1}{2n^2(r_i)\cos^2\gamma_i}\cdot\frac{\partial n^2(x, y)}{\partial y}, \qquad (2.3\text{-}10b)$$

where $r_i^2 = x_i^2 + y_i^2$.[3,10,12]

Equations (2.3-10a) and (2.3-10b) are basic ray equations that are used in the analysis of a DI rod lens.

Sometimes, as when treating a DI lens in which axial symmetry is assumed, cylindrical coordinates are more convenient for analysis. Thus we take a coordinate r in the radial direction, z in the axial direction, and θ in the angular direction, as shown in Fig. 2.3-2, where the functional is

$$F = n(r, \theta, z)\sqrt{r'^2 + (r\theta')^2 + z'^2}. \tag{2.3-11}$$

By using Euler's equations, the ray equations are rewritten in cylindrical coordinates as

$$\frac{d^2r}{d\tau^2} - r\left(\frac{d\theta}{d\tau}\right)^2 = \frac{\partial}{\partial r}\left(\frac{1}{2}n^2\right), \tag{2.3-12a}$$

$$\frac{d}{d\tau}\left(r^2\frac{d\theta}{d\tau}\right) = \frac{\partial}{\partial \theta}\left(\frac{1}{2}n^2\right), \tag{2.3-12b}$$

$$\frac{d^2z}{d\tau^2} = \frac{\partial}{\partial z}\left(\frac{1}{2}n^2\right). \tag{2.3-12c}$$

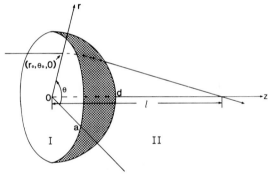

FIGURE 2.3-2. Example of ray trajectory in a 3-D DI medium. The region I is a distributed index and II a homogeneous one. (After Iga, Oikawa, and Banno.[16])

Since we choose the z axis as an optical axis, sometimes it is convenient to rewrite ray equations in other forms. By using z as an independent variable, a differential ray equation is obtained with respect to z. Here the integral of optical pass is given by

$$\mathscr{L} = \int_{P_0}^{P_1} n(r, \theta, z)\sqrt{1 + \dot{r}^2 + (r\dot{\theta})^2}\, dz, \tag{2.3-13}$$

and the associated functional is

$$F(r, \theta, \dot{r}, \dot{\theta}; z) = n(r, \theta, z)\sqrt{1 + \dot{r}^2 + (r\dot{\theta})^2}. \tag{2.3-14}$$

Euler's equations reduce to[4]

$$\ddot{r} + \frac{1}{m^2}(1 + \dot{r}^2)\dot{r}\frac{\partial}{\partial z}\left(\frac{1}{2}m^2\right) - \frac{1}{m^2}(1 + \dot{r}^2)\frac{\partial}{\partial r}\left(\frac{1}{2}m^2\right) = 0, \qquad (2.3\text{-}15)$$

$$\dot{\theta} = \frac{C\sqrt{1 + \dot{r}^2}}{mr^2}, \qquad (2.3\text{-}16)$$

where $m^2 = n^2 - C^2/r^2$, $\ddot{r} = d^2r/dz^2$, $\dot{r} = dr/dz$, $\dot{\theta} = d\theta/dz$, and C is determined by an incident condition as

$$C = n(r_0, 0)r_0^2\frac{d\theta}{ds}\bigg|_{z=0} \qquad (2.3\text{-}17)$$

The parameter m is introduced in Ref. 4. The parameter C represents the skewness of a ray. For meridional rays, $C = 0$ and $m^2 = n^2(r, z)$. We have assumed that $dn^2/d\theta = 0$.

B. RAY EQUATION IN INTEGRAL FORM

As a special case, however, there is a 3-D index distribution that is expressed by decomposed functions of radial distance r and axial distance z as shown in Eq. (2.2-1). An integral form of ray equations can be derived. These equations include the following incident conditions of the ray:

$$dr/d\tau\bigg|_{z=0} = n(r_0, 0)C_r, \qquad (2.3\text{-}18\text{a})$$

$$d\theta/d\tau\bigg|_{z=0} = n(r_0, 0)C_\theta, \qquad (2.3\text{-}18\text{b})$$

$$dz/d\tau\bigg|_{z=0} = n(r_0, 0)C_z. \qquad (2.3\text{-}18\text{c})$$

Here C_r, C_θ, and C_z are directional cosines in the cylindrical coordinate; they must satisfy the relation

$$C_r^2 + (rC_\theta)^2 + C_z^2 = 1. \qquad (2.3\text{-}19)$$

We start from Eq. (2.3-12b), and since the distributed index has axial symmetry, the partial derivative of n^2 with respect to θ is zero:

$$r^2\,d\theta/d\tau = C_0, \qquad (2.3\text{-}20)$$

where C_0 is determined by the incident condition

$$C_0 = n(r_0, 0)r_0^2 C_\theta. \qquad (2.3\text{-}21)$$

Substitution of Eq. (2.3-20) into Eq. (2.3-12a) then gives $[n(0) \equiv n(0,0)]$

$$(d^2r/d\tau^2) - (C_0^2/r^3) = \tfrac{1}{2}n^2(0)\, dR/dr. \tag{2.3-22}$$

By multiplying both sides of Eq. (2.3-22) by $dr/d\tau$, the integration may be carried out to give

$$(dr/d\tau)^2 = -(C_0/r)^2 + n^2(0)R(r) + C_1, \tag{2.3-23}$$

where C_1 is also determined by the incident condition of Eq. (2.3-18a) as

$$C_1 = n^2(r_0, 0)[C_r^2 + (rC_\theta)^2]. \tag{2.3-24}$$

From Eq. (2.3-19), we have the following relation among the constants C_r, C_θ, and C_z:

$$C_r^2 + (rC_\theta)^2 = 1 - C_z^2. \tag{2.3-25}$$

Substitution of Eq. (2.3-25) into Eq. (2.3-24) and, in turn, of Eq. (2.3-24) into Eq. (2.3-23) gives the following equation:

$$\frac{dr}{d\tau} = \pm n(0)\sqrt{\frac{-[R(r_0) + Z(0)](r_0^2 C_\theta)^2}{r^2} + R(r) - C_z^2 R(r_0) + Z(0)(1 - C_z^2)}$$

$$\tag{2.3-26}$$

($+$ for $dr/d\tau > 0$ and $-$ for $dr/d\tau < 0$). The integration of Eq. (2.3-12c) can then be carried out by multiplying both sides by $dz/d\tau$ as

$$(dz/d\tau)^2 = n^2(0)Z(z) + C_2. \tag{2.3-27}$$

The integration constant C_2 is also determined by the incident condition of Eq. (2.3-18c) as

$$C_2 = n^2(0)[(C_z^2 - 1)Z(0) + C_z^2 R(r_0)]. \tag{2.3-28}$$

Substituting Eq. (2.3-28) into Eq. (2.3-27) and considering the incident condition of the ray, that is, $dz/d\tau > 0$, we obtain

$$dz/d\tau = n(0)\sqrt{Z(z) + (C_z^2 - 1)Z(0) + C_z^2 R(r_0)}. \tag{2.3-29}$$

There may be the possibility that $dz/d\tau$ becomes negative. This may happen when the ray turns back to the incident direction.

Equations (2.3-20), (2.3-26), and (2.3-29) are first-order differential equations with respect to the parameter τ. By eliminating parameter τ the ray equations reduce to

$$\frac{d\theta}{dr} = \pm\frac{\sqrt{R(r_0) + Z(0)}\, r_0^2 C_\theta}{r^2 f(r)}, \tag{2.3-30a}$$

$$\frac{dz}{dr} = \pm\frac{\sqrt{Z(z) + (C_z^2 - 1)Z(0) + C_z^2 R(r_0)}}{f(r)}. \tag{2.3-30b}$$

These equations can be integrated, and results in the integral form are

$$\pm \int_{r_0}^{r} \frac{\sqrt{R(r_0) + Z(0)} \, r_0^2 C_\theta}{r'^2 f(r')} \, dr' = \theta - \theta_0, \tag{2.3-31a}$$

$$\int_{r_0}^{r} \frac{dr'}{f(r')} = \pm \int_{0}^{z} \frac{dz'}{\sqrt{Z(z') + (C_z^2 - 1)Z(0) + C_z^2 R(r_0)}}, \tag{2.3-31b}$$

where $f(r)$ is defined by

$$f(r) = \sqrt{\frac{-[R(r_0) + Z(0)]}{r^2}(r_0^2 C_\theta)^2 + R(r) - C_z^2 R(r_0) + Z(0)(1 - C_z^2)}. \tag{2.3-32}$$

Equations (2.3-31a) and (2.3-31b) are integral expressions of ray equations, and can also express the skewed ray trajectory.

To determine the fundamental property of the lens, the meridional ray is important. In this case the incident conditions are written in the form $C_\theta = 0$ and $C_r^2 + C_z^2 = 1$. The ray equations for the meridional ray then become

$$\int_{r_0}^{r} \frac{dr'}{\sqrt{R(r') - C_z^2 R(r_0) + Z(0)(1 - C_z^2)}}$$

$$= \pm \int_{0}^{z} \frac{dz'}{\sqrt{Z(z') + (C_z^2 - 1)Z(0) + C_z^2 R(r_0)}}. \tag{2.3-33}$$

2.4 METHOD OF RAY TRACING

In general, it is difficult to obtain an analytic solution of a ray equation in a DI medium, because the ray equation is a nonlinear differential equation. When we are required to solve a ray equation with sufficient accuracy, we have to use a numerical solution using a computer. There are some numerical methods of solving such differential equations. One simple computer method is the Runge–Kutta–Gill method, which is now prevalent in numerical analyses (see, for example, Ref. 17).

Since it is convenient to solve a ray equation by taking z as an independent variable, we use Eqs. (2.3-15) and (2.3-16), which are second-order differential equations. The examples shown here are examples for a ray tracing from a DI rod lens and DI planar microlens; the concern is with meridional rays and solving Eq. (2.3-15) in the x–z plane.

To apply the Runge–Kutta–Gill method, which approximates a differential equation to a progressive difference equation, a second-order differential equation is transferred into two first-order differential equations.

Thus, Eq. (2.3-15) is rewritten as

$$\frac{dX_1}{dz} = X_2,$$ (2.4-1a)

$$\frac{dX_2}{dz} = \frac{1}{n^2}(1 + X_2^2)X_2 \frac{\partial}{\partial z}\left(\frac{1}{2}n^2\right) - \frac{1}{n^2}(1 + X_2^2)\frac{\partial}{\partial x}\left(\frac{1}{2}n^2\right),$$ (2.4-1b)

where we have defined $X_1 = x$ and $X_2 = dx/dz$.

A. DI ROD LENS

For a DI rod lens, the partial derivative of n^2 with respect to z is zero, since the index is uniform along the z axis. By substituting Eq. (2.5-7) into Eq. (2.4-1) we obtain

$$\frac{dX_1}{dz} = X_2,$$ (2.4-2a)

$$\frac{dX_2}{dz} = \frac{n^2(0)}{n^2(X_1)}(1 + X_{2i})(-g^2 X_1 + 2h_4 g^4 X_1^3 + 3h_6 g^6 X_1^5).$$ (2.4-2b)

Equations (2.4-2a) and (2.4-2b) are solved by the Runge–Kutta–Gill method with incident conditions:[6]

$$X_{1i} = x_i,$$ (2.4-3a)

$$X_{2i} = \dot{x}_i.$$ (2.4-3b)

One problem of using the Runge–Kutta–Gill method is that we must use a suitable step-sized h. However, since it is known that the error is proportional to h^4 (we usually use the fourth-order Runge–Kutta–Gill method), by comparing solutions with different step sizes, for example, h and $h/2$, we can find a suitable step size. Figure 2.4-1 shows ray trajectories of a DI rod lens, where we have used a small enough step size of h, ($gh = 10^{-3}$ to $\sim 10^{-4}$).

B. DI PLANAR MICROLENS

In the case of a planar microlens, its index profile changes both in radial and axial direction.[18] The index profile, for example, is expressed by a quadratic form Eq. (2.2-6). Figure 2.4-2 shows an example of calculated ray trajectory of the coupled planar microlenses,[19] where Eq. (2.4-1) have been used. The employed index profile has been measured by the shearing interference method which will detailed in Chapter 8. From the measurement, the index matrix Eq. (2.2-7) is determined by the use of the least squares method. The index gradient which appears in the ray-tracing equation is obtained from the

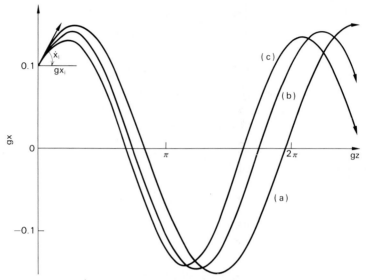

FIGURE 2.4-1.   Light ray trajectories in a DI rod lens. Eq. (2.4-2) and Eq. (2.2-2) have been used to plot the graph. The employed parameters are as follows: $gx_i = 0.1$ and $\dot{x}_i = 0.1$. (a) $h_4 = \frac{2}{3} - 5, h_6 = \frac{17}{45}$, (b) $h_4 = \frac{2}{3}, h_6 = -\frac{17}{45}$, (c) $h_4 = \frac{2}{3} + 5, h_6 = -\frac{17}{45}$.

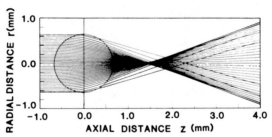

FIGURE 2.4-2.   Light ray trajectories in a planar microlens. Since these rays are all meridional, the radial distance r is used for X_1 in the text. (After Misawa, Oikawa, and Iga.[19])

ray matrix as follows:

$$\frac{\partial}{\partial x} n^2(x, z) = n^2(0)\{1, gz, (gz)^2, (gz)^3\}(N) \begin{bmatrix} 0 \\ 2g^2x \\ 4g^4x^3 \\ 6g^6x^5 \end{bmatrix}, \qquad (2.4\text{-}4a)$$

$$\frac{\partial}{\partial z} n^2(x, z) = n^2(0)\{0, g, 2g^2z, 3g^3z^2\}(N) \begin{bmatrix} 0 \\ (gx)^2 \\ (gx)^4 \\ (gx)^6 \end{bmatrix}. \qquad (2.4\text{-}4b)$$

By substituting Eqs. (2.4-4a) and (2.4-4b) into Eq. (2.4-1b), we can calculate ray trajectories using the Runge–Kutta–Gill method, as mentioned above.

2.5 DISTRIBUTED-INDEX ROD LENS

In this section we treat ray propagation in DI rod lenses. If the ray propagates in a plane that includes the optical axis, it is called a meridional ray. Since the analysis of a meridional ray is important in practical cases, we are mostly concerned here with meridional ray trajectory. Meridional rays are obtained when a ray comes from a light source on the optical axis or is incident to a lens parallel to the optical axis.

A. SOLUTION OF RAY EQUATION

Starting from Eq. (2.3-10a) and using an x–z plane as a meridional case, we first consider an approximate solution. The refractive index distribution is then taken to be

$$n^2(x) = \begin{cases} n^2(0)[1 - (gx)^2], & x \le a, \\ n_2^2, & x > a. \end{cases} \tag{2.5-1}$$

Since the index change is nearly less than 10% in practical cases, we approximate $n^2(r_i) = n^2(0)$. Moreover, we assume $1/\cos^2 \gamma_i = 1$. We then obtain

$$dx^2/dz^2 = -g^2 x. \tag{2.5-2}$$

Equation (2.5-2) gives the solution for a harmonic oscillator. By using the incident condition, that is, $x(0) = x_i$, $\dot{x}(0) = \dot{x}_i$,

$$x\Big|_{z=0} = x_i, \qquad \dot{x}\Big|_{z=0} = \dot{x}_i, \tag{2.5-3}$$

we obtain

$$x = x_i \cos gz + \frac{\dot{x}_i}{g} \sin gz. \tag{2.5-4}$$

From Eq. (2.5-4), incident rays propagate with sinusoidal curves and are focused at one point in a DI rod lens. Equation (2.5-4) is the basic solution for a DI rod lens and corresponds to paraxial rays. By using the integral form of ray equations, we can obtain an exact ray trajectory equal to Eq. (2.5-3) with suitable approximations. For the index profile of Eq. (2.5-1), Eq. (2.3-33) is written as

$$\int_{x_i}^{x} \frac{dx'}{\sqrt{1 - (gx')^2 - C_z^2\{1 - (gx_i)^2\}}} = \int_0^z \frac{dz'}{\sqrt{C_z^2\{1 - (gx_i)^2\}}}. \tag{2.5-5}$$

For simplicity, we let

$$1 - C_z^2\{1 - (gx_i)^2\} = \alpha^2$$

and

$$C_z^2\{1 - (gx_i)^2\} = 1 - \alpha^2 = \beta^2.$$

Integration of Eq. (2.5-5) results in

$$x = \frac{\alpha}{g} \sin\left(\sin^{-1}\frac{gx_i}{\alpha}\right) \cos\frac{gz}{\beta} + \frac{\alpha}{g} \cos\left(\sin^{-1}\frac{gx_i}{\alpha}\right) \sin\frac{gz}{\beta}. \qquad (2.5\text{-}6)$$

We can approximate $\alpha \simeq \dot{x}_i$, $\beta = 1$, and $\cos(\sin^{-1} gx_i/\alpha) \simeq 1$ for paraxial rays and obtain Eq. (2.5-4):

$$x = x_i \cos gz + \frac{\dot{x}_i}{g} \sin gz. \qquad (2.5\text{-}4)$$

However, the trajectory of an actual ray is slightly different from Eq. (2.5-4) when the ray is incident with an appreciable amount of offset and crosses the z axis at a large angle. In this case, we have to use more precise ray tracing.

Since an index distribution must be expressed more exactly, we represent it with higher terms:

$$n^2(x) = n^2(0)[1 - (gx)^2 + h_4(gx)^4 + h_6(gx)^6 + \cdots], \qquad (2.5\text{-}7)$$

where h_4 and h_6 are aberration coefficients of a DI rod lens. The substitution of Eq. (2.5-7) into Eq. (2.3-10a) results in

$$\frac{d^2x}{dz^2} = \frac{n^2(0)}{n^2(x_i)} (1 + \dot{x}_i)^2(-g^2x + 2h_4g^4x^3 + 3h_6g^6x^5), \qquad (2.5\text{-}8)$$

where we have used the relation

$$1/\cos^2 \gamma_i = 1 + \tan^2 \gamma_i = 1 + \dot{x}_i^2. \qquad (2.5\text{-}9)$$

An approximate solution of Eq. (2.5-8), using a perturbation theory, gives

$$x = x_i \cos \Omega z + \frac{\dot{x}_i}{\Omega} \sin \Omega z, \qquad (2.5\text{-}10)$$

where Ω is given by[10,20]

$$\Omega/g = 1 - \tfrac{3}{4}(h_4 - \tfrac{2}{3})[(gx_i)^2 + \dot{x}_i^2] - \tfrac{3}{4}(h_4 - \tfrac{2}{3})$$
$$\times [21(gx_i)^4 + 46(gx_i)^2\dot{x}_i^2 + 17\dot{x}_i^4]/12 - [\tfrac{3}{4}(h_4 - \tfrac{2}{3})]^2$$
$$\times [7(gx_i)^4 + 46(gx_i)^2\dot{x}_i^2 + 23\dot{x}_i^4]/12 - \tfrac{15}{16}(h_6 + \tfrac{17}{45})[(gx_i)^2 + \dot{x}_i^2]^2.$$
$$(2.5\text{-}11)$$

Equation (2.5-7) represents a ray trajectory that is sinusoidal with different pitches corresponding to the incident condition. Here, we should note that when

$$h_4 = \tfrac{2}{3}, \qquad h_6 = -\tfrac{17}{45}, \tag{2.5-12}$$

then $\Omega/g = 1$, that is, the pitch of each ray is independent of an incident condition. Since each ray is focused at one point in a DI rod lens, we call this an aberration-free condition. It is known that when index distribution is written as

$$n^2(x) = n^2(0)\operatorname{sech}^2(gr)$$

$$= n^2(0)[1 - (gr)^2 + \tfrac{2}{3}(gr)^4 - \tfrac{17}{45}(gr)^6 + \cdots], \tag{2.5-13}$$

a DI rod lens having no aberration for meridional rays.[21,22] The result of Eq. (2.5-12) corresponds to Eq. (2.5-13). The aberration properties of DI rod lenses are discussed in Chapter 9 using Eqs. (2.5-10) and (2.5-11).

B. NUMERICAL APERTURE OF DI ROD LENS

In this section we evaluate the output numerical aperture (NA) of a DI rod lens, which depends on the normalized index difference Δ and the length of a rod lens. The output NA is defined as the sine of the maximum angle between the output ray and the optical axis when an incident ray is parallel to the optical axis as shown in Fig. 2.5-1. Since this NA restricts the minimum focused spot size determined by diffraction, the evaluation is important when we use the rod lens as a focuser. A focused spot diameter D_s is related to NA as

$$D_s = 1.22\,\lambda/\mathrm{NA}, \tag{2.5-14}$$

where λ is wavelength.

Since a parallel ray that is incident at $x_i = a$ is

$$x(z) = a\cos gz, \qquad \dot{x}(z) = -ag\sin gz, \tag{2.5-15}$$

where a is the radius of the rod and $g = \sqrt{2\Delta}/a$, then

$$\mathrm{NA} = n_a \sin\theta_2$$

$$= n(x_0)\sin[\tan^{-1}(\sqrt{2\Delta}\sin gb)], \tag{2.5-16}$$

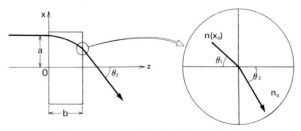

FIGURE 2.5-1. Definition of output NA.

where b is the length of the rod lens and x_0 the output position of the ray. By approximating $n(x_0) = n(0)$,

$$NA/n(0) = \sin[\tan^{-1}(\sqrt{2\Delta} \cdot \sin gb)]. \qquad (2.5\text{-}17)$$

We shall now introduce a unit $L_p = 2\pi/g$ and show NA for various index differences in Fig. 2.5-2. When a lens length is equal to $gb/2\pi = 0.25$ (i.e., $\frac{1}{4}L_p$), output NA is at a maximum. This value corresponds to incident NA, which is defined as the acceptable angle of light when the lens length is larger than $\frac{1}{4}L_p$;

$$NA_{max}/n(0) = \sin(\tan^{-1}\sqrt{2\Delta}) \approx \sqrt{2\Delta}. \qquad (2.5\text{-}18)$$

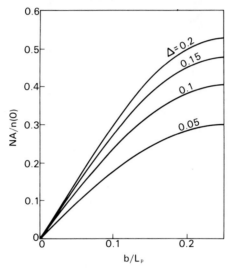

FIGURE 2.5-2.   Numerical aperture NA, where $NA = \sin\theta_2$, versus lens length of a DI rod lens, where $\Delta = \Delta n/n(0)$.

2.6 PLANAR MICROLENS WITH 3-D INDEX DISTRIBUTION

Recently the authors developed a distributed-index planar microlens.[18] The fabrication of this lens is based on the planar technology prevalent in electronics; it is made by selectively diffusing a dopant through a mask (which changes the refractive index) and into the planar substrate. It has, therefore, 3-D index distribution in both its radial and axial directions. This type of microlens has many advantages—the surface optical treatment may be easily formed in a single-batch process, many lenses (including those in a 2-D array) can be fabricated monolithically, and so on.

FIGURE 2.6-1. An interference pattern which shows equi-index surfaces of a planar microlens on the meridional plane.

An example of 3-D index distribution is shown in Fig. 2.6-1, which is an interference pattern for a longitudinal, thin plate of a planar microlens. Each interference fringe indicates an equiindex surface in the meridional plane.

To use a planar microlens, information about the ray trajectory and its dependence on lens parameters (e.g., index profile and dimensions) are necessary for lens design and also to find the optimum distribution. Here we shall try to find the fundamental ray trajectory of the planar microlens by expressing its index distribution as a decomposed function of the radial distance r and axial distance z.

We approximate a given 3-D index distribution as

$$n^2(r, z) = n^2(0, 0)[R(r) + Z(z)]. \tag{2.2-9}$$

We start from Eq. (2.3-33) and the case when rays are incident on a planar microlens parallel along the z axis. In this case, all the rays are meridional and the incident condition of Eq. (2.3-18) becomes $C_z = 1$ and $C_r = C_0 = 0$, and the ray equation (2.3-33) becomes simpler:

$$\int_{r_0}^{r} \frac{dr'}{\sqrt{R(r') - R(r_0)}} = -\int_{0}^{z} \frac{dz'}{\sqrt{Z(z') + R(r_0)}}. \tag{2.6-1}$$

The integration of both sides of Eq. (2.6-1) can be performed analytically when $R(r)$ and $Z(z)$ have simple forms.

A typical planar microlens has been made by diffusing a higher-index dopant, which brings about a higher refractive index in the substrate through circular windows in the mask.[23,24] Therefore, the planar microlens has an index distribution that decreases both in the radial and axial directions.

An index distribution is approximated as parabolic for the r direction and z directions as

$$R(r) = 1 - (gr)^2, \tag{2.6-2a}$$

$$Z(z) = -v_{20}(gz)^2. \tag{2.6-2b}$$

By substituting Eq. (2.6-2) into Eq. (2.6-1), the integration of the left-hand side of Eq. (2.6-1) becomes

$$\int_{r_0}^{r} \frac{dr'}{\sqrt{(gr_0)^2 - (gr')^2}} = \frac{1}{g}\left[\sin^{-1}\frac{r}{r_0} - \frac{(4n+1)\pi}{2}\right],$$

$$n = 0, \pm 1, \pm 2, \pm 3, \ldots. \tag{2.6-3}$$

By integrating Eq. (2.6-3) about z,[16] we can then derive:

$$r = \begin{cases} r_0 \cos\left\{\dfrac{1}{\sqrt{v_{20}}}\left[\sin^{-1}\dfrac{\sqrt{v_{20}}\,gz}{\sqrt{1-(gr_0)^2}}\right]\right\} & \text{for } v_{20} > 0. \quad (2.6\text{-}4) \\[4mm] r_0 \cos\left\{\dfrac{1}{\sqrt{-v_{20}}}\left[\sinh^{-1}\dfrac{\sqrt{-v_{20}}\,gz}{\sqrt{1-(gr_0)^2}}\right]\right\} & \text{for } v_{20} < 0. \quad (2.6\text{-}5) \end{cases}$$

By using Eq. (2.6-4), we can estimate some fundamental properties of a planar microlens.

Here, we consider a ray that is incident to a planar microlens at $(r_0, 0)$ and parallel to the z axis as shown in Fig. 2.3-2. We assume that the substrate is thick and the ray focuses in the substrate. The ray is deflected in the distributed-index region and emerges at the boundary (r_1, z_1) of the uniform-index region (where there is no dopant). We define the focusing length l which is nearly equal to focal length f, if the distributed-index region is not as thick compared with l. Since the gradient of the ray is expressed by

$$\left.\frac{dr}{dz}\right|_{z=z_1} = -\frac{\sqrt{Z(z_1) + R(r_0)}}{\sqrt{R(r_1) - R(r_0)}} \tag{2.6-6}$$

at the output position (r_1, z_1) of the distributed-index region, the focusing length l can be written as

$$l = r_1 + z_1\frac{\sqrt{Z(z_1) + R(r_0)}}{\sqrt{R(r_1) - R(r_0)}}, \tag{2.6-7}$$

where r_1 and z_1 must satisfy the following conditions:

$$\int_{r_0}^{r_1} \frac{dr}{\sqrt{R(r) - R(r_0)}} = -\int_0^{z_1} \frac{dz}{\sqrt{Z(z) + R(r_0)}}, \tag{2.6-8a}$$

$$n^2(0)[R(r_1) + Z(z_1)] = n_2^2. \tag{2.6-8b}$$

By substituting Eqs. (2.6-2a) and (2.6-2b) into Eq. (2.6-7), the focusing length l can be obtained by the equation

$$gl = \frac{gd\sqrt{2\Delta - (gr_1)^2}}{\sqrt{2\Delta}} + \sqrt{1 - 2\Delta + (gr_1)^2 - (gr_0)^2}$$

$$\times \cot\left[\frac{gd}{\sqrt{2\Delta}} \sin^{-1} \frac{\sqrt{2\Delta - (gr_1)^2}}{\sqrt{1 - (gr_0)^2}}\right], \tag{2.6-9}$$

where

$$gr_1 = gr_0 \cos\left[\frac{gd}{\sqrt{2\Delta}} \sin^{-1} \frac{\sqrt{2\Delta - (gr_1)^2}}{\sqrt{1 - (gr_0)^2}}\right]. \tag{2.6-10}$$

For a paraxial ray we can approximate $r_0 \to 0$, so the focusing length l_0 reduces to

$$gl_0 = gd + \sqrt{1 - 2\Delta} \cot\left[\frac{gd \sin^{-1} \sqrt{2\Delta}}{\sqrt{2\Delta}}\right]. \tag{2.6-11}$$

In Fig. 2.6-2, the difference in the focusing length is shown against the incident position r_0 by changing the depth of the distributed-index region d.

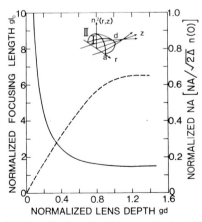

FIGURE 2.6-2. Normalized focusing length (solid curve, for $\Delta = 0.1$) and normalized numerical aperture (dashed curve) versus normalized depth of distributed-index region. I is a DI region and II a substrate. (After Iga, Oikawa, and Banno.[16])

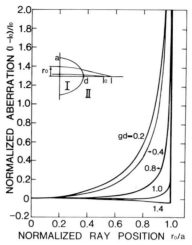

FIGURE 2.6-3.   Normalized aberration versus normalized incident position of rays at various normalized depths of the distributed-index region. I is a DI region and II a substrate. (After Iga, Oikawa, and Banno.[16])

This figure also shows the longitudinal aberration property of the assumed index distribution. The focusing length of the paraxial ray is shown in Fig. 2.6-3. We define the effective aperture r_{eff} by the radius for which the normalized longitudinal aberration $(l - l_0)/l_0$ satisfies the condition

$$(l - l_0)/l_0 < 0.5. \tag{2.6-12}$$

Then we define the effective NA, which corresponds to r_{eff} by the equation

$$\frac{\text{NA}}{n_2\sqrt{2\Delta}} = \frac{1}{gl_0}\left(\frac{r_{\text{eff}}}{a}\right), \tag{2.6-13}$$

where the effective NA is normalized by $n_2\sqrt{2\Delta}$. The normalized NA is also shown in Fig. 2.6-4 as a function of the depth of the distributed-index region. Here we define NA as a sine of the maximum angle of the ray which comes into the region of $(l - l_0)/l_0 = \varepsilon$. The normalized NA has a peak corresponding to the depth of the distributed-index region if we take the allowance of the aberration smaller than $(l - l_0)/l_0 = 0.01$ to define the effective aperture. This result means that the aberration can be reduced by choosing a suitable axial-index distribution.

A numerical example of a typical planar microlens made by the electromigration technique might have $\Delta = 0.05$, $d = 0.4$ mm, $a = 0.5$ mm, $n_2 = 1.5$, $g = \sqrt{2\Delta}/a = 0.63$ mm^{-1}, $gd = 0.25$, $l_0 = 3.5/g = 5.56$ mm, and NA $= 0.2\sqrt{2\Delta}n_2 = 0.09$.

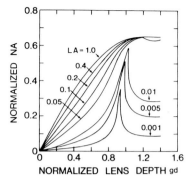

FIGURE. 2.6-4. Normalized numerical aperture versus normalized depth of distributed-index region at various normalized longitudinal aberrations LA. (After Iga, Oikawa, and Banno.[16])

The focusing length l and NA can be compared with the experimental data of a planar microlens made by the electromigration technique,[24] that is, $l = 5.17$ mm (in glass) and NA $= 0.15$, showing good agreement between l's. The larger NA of the experimental value is due to the smaller aberration of the actual planar microlens.

When we assume that we can take larger Δn, for example, $\Delta n = 0.23$, $d = 0.4$ mm, $a = 0.5$ mm, and $n_2 = 1.6$, we can obtain $g = \sqrt{2\Delta}/a = 1.02$ mm^{-1}, $gd = 0.41$, $l_0 = 2.5/g = 2.45$ mm, and NA $= 0.2\sqrt{2\Delta}n_2 = 0.23$.

2.7 THE WAVEFRONT AND LIGHTRAY CONCEPTS

The lightray we have introduced is convenient and easy to understand. However, when we attempt to solve a diffraction problem, we should treat the light as a wave. By using a wavefront having a certain relation to the lightray, we can accomplish this.

The wavefront is defined as a surface constructed with a set of equioptical passlength points from a light source. On a wavefront the light phase is constant. The wavefront is described by the eikonal equation[2]

$$\left(\frac{\partial \varphi}{\partial x}\right)^2 + \left(\frac{\partial \varphi}{\partial y}\right)^2 + \left(\frac{\partial \varphi}{\partial z}\right)^2 = n^2(x, y, z), \tag{2.7-1}$$

where φ is the light phase and corresponds to the optical pass. The wavefront is defined as $\varphi = $ const.

By using the wavefront concept, rays are also defined as curves that are orthogonal to the wavefront, as shown in Fig. 2.7-1. Since the $\nabla \varphi$ is orthogonal to the wavefronts $\varphi = $ const., the ray equation can be written using φ:

$$dx/d\tau = \dot{\partial}\varphi/\partial x, \tag{2.7-2a}$$

$$dy/d\tau = \partial\varphi/\partial y, \tag{2.7-2b}$$

$$dz/d\tau = \partial\varphi/\partial z, \tag{2.7-2c}$$

where, as before, $d\tau = ds/n$. By differentiating Eq. (2.7-2a) with respect to τ, to obtain

$$\begin{aligned}
\frac{d^2x}{d\tau^2} &= \frac{\partial^2\varphi}{\partial x^2}\frac{dx}{d\tau} + \frac{\partial^2\varphi}{\partial x\,\partial y}\frac{dy}{d\tau} + \frac{\partial^2\varphi}{\partial x\,\partial z}\frac{dz}{d\tau} \\
&= \frac{\partial^2\varphi}{\partial x^2}\frac{\partial\varphi}{\partial x} + \frac{\partial^2\varphi}{\partial x\,\partial y}\frac{\partial\varphi}{\partial y} + \frac{\partial^2\varphi}{\partial x\,\partial z}\frac{\partial\varphi}{\partial z} \\
&= \frac{1}{2}\frac{\partial}{\partial x}\left[\left(\frac{\partial\varphi}{\partial x}\right)^2 + \left(\frac{\partial\varphi}{\partial y}\right)^2 + \left(\frac{\partial\varphi}{\partial z}\right)^2\right] \\
&= \frac{1}{2}\frac{\partial}{\partial x}n^2.
\end{aligned} \tag{2.7-3a}$$

Similarly, by differentiating Eqs. (2.7-2b) and (2.7-2c) with respect to τ, we obtain

$$\frac{d^2y}{d\tau^2} = \frac{\partial}{\partial y}\left(\frac{1}{2}n^2\right), \tag{2.7-3b}$$

$$\frac{d^2z}{d\tau^2} = \frac{\partial}{\partial y}\left(\frac{1}{2}n^2\right). \tag{2.7-3c}$$

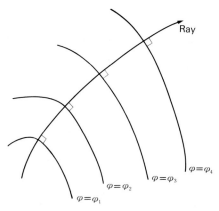

Wavefronts

FIGURE 2.7-1.   Relation between wavefronts and lightray.

These equations correspond to Eq. (2.3-7), which was obtained from Fermat's principle.

A. PLANE WAVE AND PARALLEL RAYS

If we think of parallel rays that progress in the (k_x, k_y, k_z) direction, by definition, a wavefront is a plane surface as shown in Fig. 2.7-2. We shall now reduce the wavefront equations that progress in the **k** direction. At the point $P(x, y, z)$ phase is

$$\varphi(x, y, z) = \overline{O'P}(2\pi/\lambda) + \varphi_0, \tag{2.7-4}$$

where φ_0 is the phase of the light at the origin and λ is wavelength, since $\overline{O'P}$ is obtained from a scalar product of **k** and **r**,

$$\varphi(x, y, z) = \hat{\mathbf{k}} \cdot \mathbf{r}(2\pi/\lambda) + \varphi_0 = \mathbf{k} \cdot \mathbf{r} + \varphi_0, \tag{2.7-5}$$

where $\hat{\mathbf{k}} = \mathbf{k}/|\mathbf{k}|$. Then we can describe the plane as

$$u = u_0 \exp[\,j(\omega t - \mathbf{k} \cdot \mathbf{r} - \varphi_0)]$$

If wavefront is

$$\mathbf{k} \cdot \mathbf{r} + \varphi_0 = C(\text{const.}), \tag{2.7-6}$$

then

$$k_x x + k_y y + k_z z = C - \varphi_0 \tag{2.7-7}$$

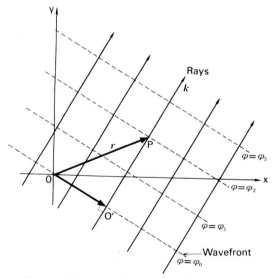

FIGURE 2.7-2.   Parallel rays and a plane wavefront.

B. Focusing Rays and Spherical Wavefronts

When the light rays are focused at one point 0, a set of the equioptical length from the origin O is a sphere (Fig. 2.7-3). The sphere is orthogonal to the convergent rays, and a lens wave aberration is defined from this principle. The lens is used to focus the lightrays on one point, but when the lens has an aberration, the rays cannot focus. In this case, the wavefront is nonspherical and the displacement of the actual wavefront from a spherical one is the wave aberration. Since ray aberration is also reduced from a wave aberration using Eq. (2.7-2), wave aberration is measured to evaluate lenses, as shown in Fig. 2.7-3.

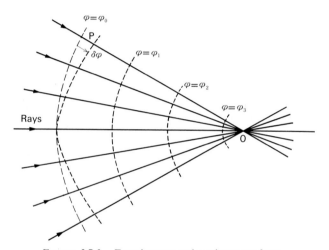

FIGURE. 2.7-3.   Focusing rays and a sphere wavefront.

REFERENCES

1.   R. W. Wood, *Physical Optics*. Macmillan, New York, 1905.
2.   M. Born and E. Wolf, *Principle of Optics*. Pergamon, Oxford, 1959.
3.   R. K. Luneburg, *Mathematical Theory of Optics*. Univ. of California Press, Berkeley, 1964.
4.   E. W. Marchand, *Gradient Index Optics*. Academic Press, New York, 1978.
5.   D. T. Moore, *J. Opt. Soc. Am.* **65**, 451 (1975).
6.   K. Iga, *Appl. Opt.* **19**, 1039 (1982).
7.   Y. Suematsu and K. Iga, *Trans. IECE Jpn.* **49**, 1645 (1966) (in Japanese) [an English version is available from Scripta Electronica Japonica, New York].
8.   K. Iga, K. Yokomori, and T. Sakayori, *Appl. Phys. Lett.* **26**, 578 (1975).
9.   Y. Suematsu and K. Furuya, *Trans. IECE Jpn.* **B54**, 325 (1971).
10.   K. Iga, *Appl. Opt.* **19**, 1039 (1980).
11.   R. K. Luneburg, *Mathematical Theory of Optics*. Univ. California Press, Berkeley, 1964.
12.   W. Streifer and K. B. Paxton, *Appl. Opt.* **10**, 769 (1971).
13.   D. Gloge and E. A. J. Marcatili, *Bell Syst. Tech. J.* **52**, 1563 (1973).

14. I. Kitano, K. Koizumi, H. Matsumura, T. Uchida, and M. Furukawa, *Jpn. J. Appl. Phys. Suppl.* **39**, 63 (1970).
15. M. Oikawa, K. Iga, and T. Sanada, *Jpn. J. Appl. Phys.* **20**, L51 (1981).
16. K. Iga, M. Oikawa, and J. Banno, *Appl. Opt.* **21**, 3451 (1982).
17. F. B. Hildebrand, *Introduction to Numerical Analysis*. McGraw-Hill, New York, 1973.
18. M. Oikawa and K. Iga, *Appl. Opt.* **21**, 1052 (1982).
19. S. Misawa, M. Oikawa, and K. Iga, 4th Topical Meeting on Gradient-Index Optical Imaging Systems, No. D2, 1983.
20. N. Yamamoto and K. Iga, *Appl. Opt.* **19**, 1101 (1980).
21. K. Unger, *Annalen der Physik*, **19**, 64 (1967).
22. S. Kawakami and J. Nishizawa, *IEEE Trans. Microwave Theory Tech.* **MTT-16**, 814 (1968).
23. M. Oikawa, K. Iga, T. Sanada, N. Yamamoto, and K. Nishizawa, *Jpn. J. Appl. Phys.* **20**, L296 (1981).
24. M. Oikawa, K. Iga, and T. Sanada, *Electron. Lett.* **17**, 452 (1981).

CHAPTER 3

Imaging and Focusing— Ray Theory

3.1 OPENING REMARKS

Recently, fibers and lenses with radially distributed (square-law) index profiles have been adapted to a wide range of applications including imaging systems,[1,2] focusing devices,[3] coupling lenses,[4] and microoptical circuits.[5,6] A Wood lens[7] is one of these, and Mikaeliyan[8] has presented their ray-focusing property. The first such device made was a gas lens proposed and demonstrated by Berreman.[9] Some gas lenses have been studied as light transmission lines,[10,11] and Hermite–Gaussian[12] or Laguerre–Gaussian modes[13] have been introduced as the normal mode in such a lenslike medium. A remarkable aberration has been observed in laser beam transmission through a flow-type gas lens, and a perturbation method has been introduced to solve this problem.[14] Imaging with a gas lens has also been demonstrated.[15]

In 1969 light-focusing fiber[16] and graded-index lenses[17] were made with glass materials. The optimum index distribution for reducing the delay distortion of a light pulse was obtained theoretically,[18] and since that time there has been a lot of work on widening the bandwidth of multimode graded-index fibers.[19–21]

On the other hand, the fundamental concepts of imaging with a distributed-index or gradient-index medium have been presented,[16,22] and considerations on aberrations given.[23] It has been recognized that improvement of the fabrication process is one of the most important problems that needs to be tackled if aberrations are to be reduced.[3,24] To feed back information on the optical property of a DI lens to the fabrication process, a simple and self-consistent theory exhibiting wave–optical and ray–optical properties is necessary, as well as precise measuring techniques.[25,26]

In this chapter we present a simple and convenient expression for DI imaging to assist in analyzing the aberrations of imaging systems by applying ray optics.[27] Some criteria and design considerations will be given for applying this theory.

3.2 SIMPLE EXPLANATION OF DI IMAGING

As shown in Chapter 2, ray transmission in a DI rod that has a parabolic index distribution is expressed simply by ray equations[28,29].

The ray transmitted along the axis obeys ray equations (2.2-4a) and (2.2-4b). If we look at a meridional ray in the $x-z$ plane, the solution of ray equation (2.2-4a) is given by

$$x = x_i \cos \Omega z + \dot{x}_i/g \sin \Omega z, \tag{3.2-1}$$

$$\frac{\Omega}{g} = 1 - \frac{3}{4}\left(h_4 - \frac{2}{3}\right)[(gx_i)^2 + \dot{x}_i^2] + \left(h_4 - \frac{2}{3}\right)O(x_i^4) + \left(h_6 + \frac{17}{45}\right)O(x_i^6), \tag{3.2-2}$$

where $\dot{x}_i$ is the ray slope at $z = 0$. This is derived using the standard method of solving a nonlinear equation.[29] The explicit formula to the sixth order can be found in Reference 25 and Chapter 9 of this text. The validity of the equation has been confirmed by comparing the numerical solution of the ray equation.[30]

When we take the first term of (3.2-2) and neglect the higher-order terms, we obtain

$$x = x_i \cos(gz) + (\dot{x}_i/g) \sin(gz). \tag{3.2-3}$$

By differentiating (3.2-2) with respect to z we obtain a ray with slope x inside the rod:

$$\dot{x} = -gx_i \sin(gz) + \dot{x}_i \cos(gz). \tag{3.2-4}$$

The trace of two representative rays coming from an object is shown in Fig. 3.2-1, where we do not take refraction at both surfaces into account.

We show in Fig. 3.2-2 some imaging configurations. The first case is where the length L of the rod is such that $L = \frac{1}{4}L_p$, where the pitch length L_p is given by $2\pi/g$. Here we can see the rod is imaging as a single convex lens. The second case is the $\frac{1}{2}L_p$ configuration. In this case the rays are emit at spreading angles and a real image is no longer formed outside the rod.

Instead, a virtual image is observed as shown in the figure. When the length is $\frac{3}{4}L_p$, the real image is formed. This is one of the specific characteristics of DI imaging, that is, if we use another piece of rod close to the first, the image made by the second lens can be overlapped to the first. This concept is equivalent to a relay of lenses to form an elect image and is actually utilized in making a conjugate image in copy machines and facsimiles. In similar ways, the one pitch length rod forms an elect virtual image as shown in Fig. 3.2-2d. When the length of the rod is longer than L_p, the image is relayed as shown in Fig. 3.2-2.

The magnification of the image is dependent on the distance of the object from the surface of the rod and, of course, on the rod parameter. This will be detailed in a latter chapter by the use of the ray matrix method.

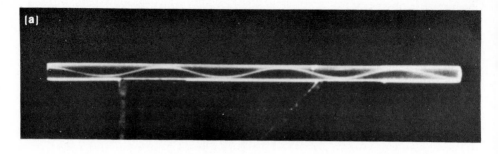

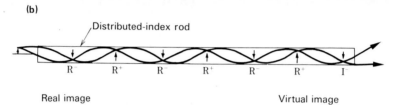

Real image   Virtual image

FIGURE 3.2-1.   (a) A sinusoidal light beam trajectory in a distributed-index rod lens. The rod diameter is 3 mm. (b) A simplitied drawing showing how an image is transmitted through the distributed-index rod.

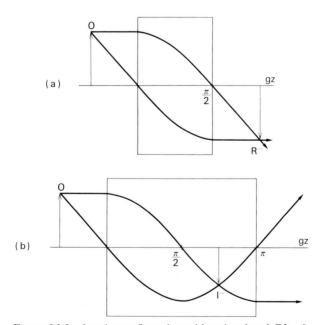

FIGURE 3.2-2.   Imaging configurations with various length DI rods.

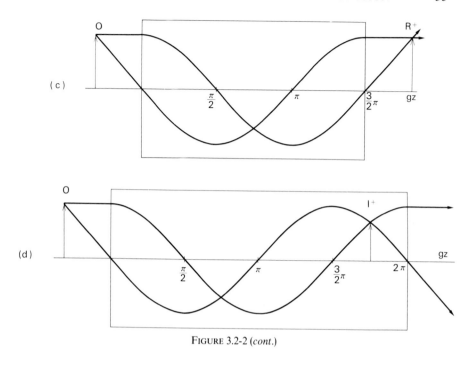

FIGURE 3.2-2 *(cont.)*

3.3 SPOT DIAGRAM AND OPTICAL
TRANSFER FUNCTION

The rays incident parallel to the axis are focused at $z = (\pi/g)(2p + 1)$ where p is an integer. The dispersion of the ray position at the focal point is expressed in terms of a spot diagram, which is easily obtained if we use Eqs. (3.2-1) and (3.2-2). Let us consider the spot that comes from point source P shown in Fig. 3.3-1, where we have assumed that this system makes a real image with unit magnification. The associated spot diagram is shown in Fig. 3.3-2. We shall use the notation J as a measure of the height x_0 given by $x_0 = (J/5)\sqrt{2} A$, where A is a radius of the DI medium and $x_0 = \sqrt{2} A$ corresponds to the highest object position included in the NA of the lens. We have also assumed that $g = 0.17$ mm^{-1}, $A = 1$ mm, and $n(0) = 1.5$.

The optical transfer function (OTF) $H(u)$, here u is the spatial frequences, can be quickly obtained from the spot diagram. The magnitude of the OTF $T(u')$ is shown in Fig. 3.3-3, where u' is a normalized spatial frequency defined by

$$u' = [u/gn(0)][\tfrac{3}{4}(1 + gb)|h_4 - \tfrac{2}{3}|n^2(0)].\qquad(3.3\text{-}1)$$

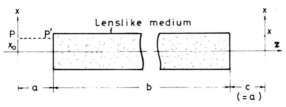

FIGURE 3.3-1.   Configuration for making a spot diagram, where x is traverse position, a and b are lengths of free space and rod, respectively. The point P is the object, and P' denotes the point that has the same height as P. (After Iga, Hata, Kato, and Fukuyo;[31] Iga.[32])

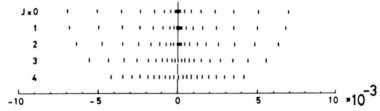

FIGURE 3.3-2.   One-dimensional spot diagram. The horizontal axis denotes the relative ray position given by $gn^3(0)(x - x_0)/[\frac{-3}{4}(1 + gb)(h_4 - \frac{2}{3})]$. (After Iga, Hata, Kato, and Fukuyo.[31])

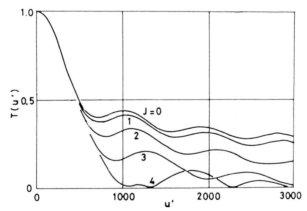

FIGURE 3.3-3.   Amplitude of OTF. (After Iga, Hata, Kato, Fukuyo;[31] Iga.[32])

When $T(u') = \frac{1}{2}$, bandwidth B of the transmitted spatial frequency can be given as

$$B = 460gn^3(0)/[\tfrac{3}{4}(1 + gb)|h_4 - \tfrac{2}{3}|].\qquad(3.3\text{-}2)$$

The bandwidth is inversely proportional to $h_4 - \frac{2}{3}$ to the order. If we assume $g = 0.17$ mm^{-1}, $n(0) = 1.5$, $gb = \pi/2$, and $h_4 - \frac{2}{3} = 0.1$, B is nearly equal to 100 mm^{-1}.

3.4 FOCUSING

A DI lens may be applied as a light focuser when the focal point is made at the end or outside of a lens piece by cutting a sample with nearly a $\frac{1}{4}$ pitch. In Fig. 3.4-1 we show two examples of ray traces focused by a lens with aberration. It is, therefore, necessary to reduce the aberration originating from higher-order terms in index profile by optimization of fabrication process. The intensity distribution of a focused spot is shown in Fig. 3.4-2.

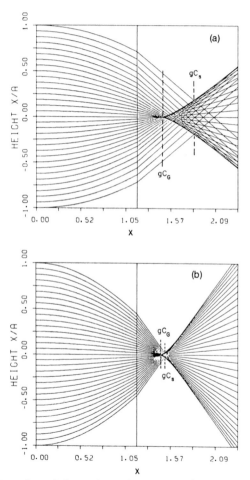

FIGURE 3.4-1. Rays focused by a lens shorter than $\frac{1}{4}$ pitch when (a) $h_4 - \frac{2}{3} = 10$; (b) $h_4 - \frac{2}{3} = 2$. The horizontal axis denotes the x axis normalized by g. The planes C_G and C_S are the Gaussian image plane and the circle of least confusion, respectively. (After Iga, 1980.)

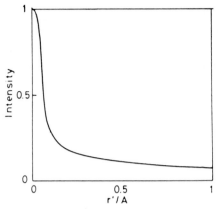

FIGURE 3.4-2.   Intensity distribution of a focused spot (ray optics). (After Iga.[32])

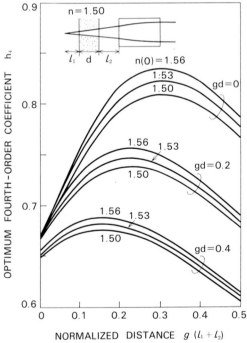

FIGURE 3.4-3.   Optimum h_4 for focusing. (After Iga.[32])

When we want to focus light at the back surface of a transparent parallel plate as shown in Fig. 3.4-3, we can choose the fourth-order coefficient h_4 to compensate for the aberration due to refraction at the plane surface.[30] The optimum h_4 is calculated as shown in Fig. 3.4-3 as a function of lens length and spacing.

REFERENCES

1. M. Kawazu and Y. Ogura, in Digest of Topical Meeting on Gradient Index Optical Imaging Systems, Paper WC1. Optical Society of America, Washington, D.C., 1979.
2. M. Mino, in Ref. 1, paper WC3.
3. T. Miyazawa, K. Okada, T. Kubo, K. Nishizawa, I. Kitano, and K. Iga, in Ref. 1, Paper WA3.
4. K. Kobayashi, R. Ishikawa, K. Minemura, and S. Sugimoto, in Technical Digest, International Conference on Integrated Optics and Optical Fiber Communication, Tokyo (1977), p. 367.
5. K. Kobayashi, R. Ishikawa, K. Minemura, and S. Sugimoto, *Fiber Integr. Opt.* **2**, 1 (1979).
6. W. J. Tomlinson, in Digest of Topical Meeting on Gradient Index Optical Imaging Systems, Paper WC6. Optical Society of America, Washington, D.C. 1979.
7. R. W. Wood, *Physical Optics.* Macmillan, New York, 1905.
8. A. L. Mikaeliyan, *Dokl. Akad. Nauk SSSR*, **81**, 569 (1951).
9. D. W. Berreman, *Bell Syst. Tech. J.* **43**, 1469 (1964).
10. Y. Suematsu, K. Iga, and S. Ito, *IEEE Trans. Microwave Theory Tech.* **MTT-14**, 657 (1966).
11. P. Kaiser, *Bell Syst. Tech. J.* **49**, 137 (1970).
12. E. A. J. Marcatili, *Bell Syst. Tech. J.* **43**, 2887 (1964).
13. Y. Suematsu and H. Fukinuki, *J. IECE Jpn.* **48**, 1684 (1965).
14. Y. Suematsu and K. Iga, *J. IECE Jpn.* **49**, 1645 (1966).
15. Y. Aoki and M. Suzuki, *IEEE Trans. Microwave Theory and Tech.* **MTT-15**, 1 (1967).
16. T. Uchida, M. Furukawa, I. Kitano K. Koizumi, and H. Matsumura, *IEEE J. Quantum Electron.* **QE-6**, 606 (1970).
17. A. D. Pearson, W. G. French, and E. G. Rawson, *Appl. Phys. Lett.* **15**, 76 (1969).
18. S. Kawakami and J. Nishizawa, *IEEE Trans. Microwave Theory Tech.* **MTT-16**, 814 (1968).
19. D. Glóge, *Appl. Opt.* **10**, 2442 (1971).
20. R. Olshansky and D. B. Keck, *Appl. Opt.* **15**, 483 (1976).
21. K. Furuya, Y. Suematsu, J. Nayyer, and S. Ishikawa, *Appl. Phys. Lett.* **27**, 456 (1975).
22. E. W. Marchand, *Gradient Index Optics.* Academic Press New York, 1978.
23. A. K. Gathak and K. Thyagarajan, *Contemporary Optics.* Plenum, New York, 1978.
24. K. Iga and N. Yamamoto, *Appl. Opt.* **16**, 1305 (1977).
25. N. Yamamoto and K. Iga, in Digest of Topical Meeting on Gradient Index Optical Imaging Systems, Paper WC3. Optical Society of America, Washington, D.C., 1979.
26. K. Iga and Y. Kokubun, *Appl. Opt.* **17**, 1972 (1978).
27. K. Iga, in Digest of Topical Meeting on Gradient Index Optical Imaging Systems, Paper TuA2. Optical Society of America, Washington, D.C., 1979.
28. R. K. Luneburg, *Mathematical Theory of Optics.* Univ. of California Press, Berkeley, 1964.
29. W. Streifer and K. B. Paxton, *Appl. Opt.* **10**, 769 (1971).
30. K. Iga and S. Ohshima, unpublished work; M. A. Thesis, Tokyo Institute of Technology (1978).
31. K. Iga, S. Hata, Y. Kato, and H. Fukuyo, *Jpn. J. Appl. Phys.*, **13**, 79 (1974).
32. K. Iga, *Appl. Opt.* **19**, 1039 (1980).

CHAPTER 4

Guided-Wave Theory

4.1 OPENING REMARKS

Ray optics provides a simple and straightforward way of understanding the focusing and imaging properties of a DI system. But it is almost useless in describing diffraction, and wave optics is necessary in many situations. Of course, we can artificially include the diffraction term in a ray equation, such as the paraxial ray equation

$$d^2x/dz^2 = g^2x. \tag{4.1-1}$$

Here, we have assumed coefficient $A = 1$, and only the parabolic term of Eq. (2.2-2) is taken into account. If we introduce the term $-1/k^2(0)x^3$ into the right-hand side of Eq. (4.1-1)[1], we have

$$d^2x/dz^2 = g^2x - 1/k^2(0)x^3, \tag{4.1-2}$$

where $k(0) = 2\pi n(0)/\lambda$ and $n(0)$ is the refractive index at the center axis. Then we can set $d^2x/dz^2 = 0$, and the corresponding ray position x_0 is given by

$$x_0 = 1/\sqrt{gk(0)}. \tag{4.1-3}$$

As we shall see from wave theory results, x_0 can be attributed to the characteristic spotsize of the Gaussian mode for a parabolic DI fiber.

Wave theory is not only effective for dealing rigorously with diffraction, as we have just seen, but can also express coherent image transmission through DI fibers. In this chapter, we shall describe the propagation characteristics of some typical optical waveguides, such as step-index and distributed-index slab waveguides and round optical fibers. The slab waveguide will be used in many optical components, such as semiconductor lasers and integrated optics.

Although the temporal pulse dispersion characteristic is important for optical fibers as an information transmission media, we shall concentrate here on describing spatial waveguiding phenomena, because the aim of this book is to provide readers with the fundamental understanding of optical fibers as a part of microoptics.

4.2 STEP-INDEX PLANAR WAVEGUIDE

First, we start with derivating wave equations for a dielectric slab waveguide. Assuming a time and distance dependence of $\exp[j(\omega t - \beta z)]$, Maxwell's equations are

$$\mathbf{V} \times \mathbf{E} = -j\omega\mu\mathbf{H}, \tag{4.2-1}$$

$$\mathbf{V} \times \mathbf{H} = j\omega\varepsilon\mathbf{E}, \tag{4.2-2}$$

$$\mathbf{V} \cdot \mathbf{D} = 0, \tag{4.2-3}$$

$$\mathbf{V} \cdot \mathbf{B} = 0. \tag{4.2-4}$$

With the help of a vector formula, we have

$$\mathbf{V} \times \mathbf{V} \times \mathbf{E} = \mathbf{V} \cdot (\mathbf{V} \cdot \mathbf{E}) - \mathbf{V}^2\mathbf{E}, \tag{4.2-5}$$

where $\mathbf{V}^2$ denotes the Laplacian operator. Substituting Eqs. (4.2-1) and (4.2-3) into (4.2-5), a wave equation for $\mathbf{E}$ can be obtain:

$$\mathbf{V}^2\mathbf{E} + \omega^2\varepsilon\mu\mathbf{E} = -\mathbf{V}\left(\mathbf{E} \cdot \frac{\mathbf{V}\varepsilon}{\varepsilon}\right). \tag{4.2-6}$$

A similar wave equation is derived from Eqs. (4.2-2) and (4.2-4) in the same way:

$$\mathbf{V}^2\mathbf{H} + \omega^2\varepsilon\mu\mathbf{H} = -\frac{\mathbf{V}\varepsilon}{\varepsilon} \times (\mathbf{V} \times \mathbf{H}). \tag{4.2-7}$$

Now we assume a symmetric slab waveguide as shown in Fig. 4.2-1. The refractive indices of the core and cladding are designated as n_1 and n_2, respectively, and the thickness of the core is $2a$. In the present case, since $\mathbf{V}\varepsilon = 0$, we obtain the next wave equations for the E_z and H_z components:

$$\frac{\partial^2 E_z}{\partial x^2} + (k_0^2 n^2 - \beta^2)E_z = 0, \tag{4.2-8}$$

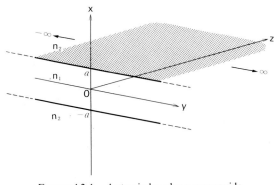

FIGURE 4.2-1. A step-index planar waveguide.

$$\frac{\partial^2 H_z}{\partial x^2} + (k_0^2 n^2 - \beta^2)H_z = 0, \tag{4.2-9}$$

where $\mu = \mu_0$, $n^2 = \varepsilon/\varepsilon_0$ and $k_0^2 = \omega^2 \mu_0 \varepsilon_0$.

The transverse field components are expressed in terms of E_z and H_z with the aid of Eqs. (4.2-1) and (4.2-2) as follows:

$$E_x = -\frac{j}{\omega^2 \varepsilon \mu - \beta^2}\left(\beta \frac{\partial E_z}{\partial x} + \omega\mu \frac{\partial H_z}{\partial y}\right), \tag{4.2-10}$$

$$E_y = \frac{j}{\omega^2 \varepsilon \mu - \beta^2}\left(-\beta \frac{\partial E_z}{\partial y} + \omega\mu \frac{\partial H_z}{\partial x}\right), \tag{4.2-11}$$

$$H_x = \frac{j}{\omega^2 \varepsilon \mu - \beta^2}\left(\omega\varepsilon \frac{\partial E_z}{\partial y} - \beta \frac{\partial H_z}{\partial x}\right), \tag{4.2-12}$$

$$H_y = \frac{-j}{\omega^2 \varepsilon \mu - \beta^2}\left(\omega\varepsilon \frac{\partial E_z}{\partial x} + \beta \frac{\partial H_z}{\partial y}\right). \tag{4.2-13}$$

The guided modes must satisfy Eqs. (4.2-8) and (4.2-9). However, because these equations are independent of each other, general solutions of Eqs. (4.2-8) and (4.2-9) can be expressed by the sum of two independent solutions, that is, one satisfies Eq. (4.2-8) with $H_z = 0$ and the other Eq. (4.2-9) with $E_z = 0$. The latter solutions are called TE modes, because only the transverse electric field E_y is not zero. In the same way, the former ones are called TM modes. From $\partial/\partial y = 0$, the following field components remain:

TE modes $E(0, E_y, 0)$, $H(H_x, 0, H_z)$,

TM modes $E(E_x, 0, E_z)$, $H(0, H_y, 0)$.

A. TE MODES

The tangential field components of TE modes are E_y and H_z. Since these two are related directly to the boundary condition, it is convenient to deal with the E_y component instead of E_z. From Eq. (4.2-6), E_y must satisfy

$$\frac{\partial^2 E_y}{\partial x^2} + (n_1^2 k_0^2 - \beta^2)E_y = 0 \qquad \text{(core)} \tag{4.2-14}$$

$$\frac{\partial^2 E_y}{\partial x^2} - (\beta^2 - n_2^2 k_0^2)E_y = 0 \qquad \text{(cladding)}. \tag{4.2-15}$$

Special solutions of Eq. (4.2-14) in the core are cosine and sine functions. In the cladding, the solutions are classified into two types, namely the evanescent (exponentially decaying) solution for $n_2 k < |\beta| < n_1 k$ and the sinusoidal oscillating solution for $|\beta| < n_2 k$. The former is called the guided modes. Some amount of optical power of a guided mode is confined in the core, and the remainder permeates from the cladding. The latter is called a set of radiation modes and the power is no more confined in the core. The group of all the guided and radiation modes constitutes a complete orthogonal set and any field can be expanded in terms of these guided and radiation modes. At first, we discuss the guided modes.

The solution of a guided mode must satisfy the boundary conditions stating that the tangential components of the electric field must be continuous at the core–cladding boundary and approach zero at $x \to \infty$. From the above conditions, we obtain the following mode distributions:

Even modes $\quad E_y = \begin{cases} A_e \cos(\kappa x), & |x| \leq a, \\ A_e \cos(\kappa a) \exp[-\gamma(|x| - a)], & |x| > a; \end{cases}$ (4.2-16)

Odd modes $\quad E_y = \begin{cases} A_0 \sin(\kappa x), & |x| \leq a, \\ (x/|x|) A_0 \sin(\kappa a) \exp[-\gamma(|x| - a)] & |x| > a; \end{cases}$

(4.2-17)

where

$$\kappa^2 = k_0^2 n_1^2 - \beta^2,$$ (4.2-18)

$$\gamma^2 = \beta^2 - k_0^2 n_2^2.$$ (4.2-19)

Then H_z is another tangential field component continuous at the core–cladding boundary. From this boundary condition, the eigenvalue equations for TE even and odd modes are derived as Eqs. (4.2-20) and (4.2-21), respectively, as tabulated in Table 4.2-I. The solutions of these eigenvalue equations can be normalized by introducing new parameters b and V, defined by

$$b = \gamma^2/(\kappa^2 + \gamma^2) = \frac{(\beta/k_0)^2 - n_2^2}{(n_1^2 - n_2^2)},$$ (4.2-22)

$$V = [(\kappa a)^2 + (\gamma a)^2]^{1/2} = k_0 n_1 a \sqrt{2\Delta},$$ (4.2-23)

where

$$2\Delta = \frac{n_1^2 - n_2^2}{n_1^2} \simeq 2\frac{n_1 - n_2}{n_1}$$ (4.2-24)

and giving a dispersion curve as shown in Fig. 4.2-2. When the waveguide parameters n_1, n_2, and a and wavelength λ are given, the propagation constant β of any mode can be obtained from Fig. 4.2-2 through Eqs. (4.2-22) and (4.2-23). The mode number is labeled in the order of decreasing β, and TE_0 is the fundamental mode. This mode number corresponds to the number of nodes in the field distribution.

TABLE 4.2-I

MODE FUNCTIONS OF STEP-INDEX PLANAR WAVEGUIDES

Mode	Mode distribution		Eigenvalue equation[a]
	for $|x| \leqq a$	for $|x| > a$	
TE (even)	$E_y = A_e \cos \kappa x$	$E_y = A_e \cos(\kappa a) e^{-\gamma(|x|-a)}$	$\tan \kappa a = \dfrac{\gamma a}{\kappa a}$ $\qquad$ (4.2-20)
TE (odd)	$E_y = A_0 \sin \kappa x$	$E_y = \dfrac{x}{|x|} A_0 \sin(\kappa a)\, e^{-\gamma(|x|-a)}$	$\tan(\kappa a) = -\dfrac{\kappa a}{\gamma a}$ $\qquad$ (4.2-21)
TM (even)	$H_y = B_e \cos \kappa x$	$H_y = B_e \cos(\kappa a)\, e^{-\gamma(|x|-a)}$	$\tan \kappa a = \left(\dfrac{n_1}{n_2}\right)^2 \dfrac{\gamma a}{\kappa a}$ $\qquad$ (4.2-32)
TM (odd)	$H_y = B_e \sin \kappa x$	$H_y = \dfrac{x}{|x|} B_0 \sin(\kappa a)\, e^{-\gamma(|x|-a)}$	$\tan \kappa a = -\left(\dfrac{n_2}{n_1}\right)^2 \dfrac{\kappa a}{\gamma a}$ $\qquad$ (4.2-33)

[a] The solutions of the eigenvalue equations can be normalized using V, where $V^2 = (\kappa a)^2 + (\gamma a)^2$.

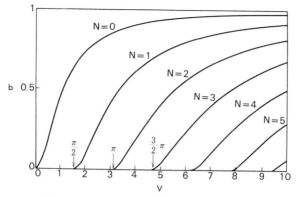

FIGURE 4.2-2. Dispersion curves for TE modes.

When the propagation constant of one guided mode reaches $n_2 k \, (b \rightarrow 0)$, the mode is "cutoff," and the V value is then called the "cutoff V value." Putting $\gamma = 0$ and $\kappa a = V$, the cutoff V value of TE_N modes is easily obtained from Eqs. (4.2-18)–(4.2-21) as

$$.V_C = N(\pi/2), \qquad N = 0, 1, 2, \ldots. \qquad (4.2\text{-}25)$$

The cutoff V value of TE_1 gives a single-mode condition, because, when V is smaller than $\pi/2$, only the TE_0 mode can propagate. The single-mode condition is important for designing single-mode waveguides and single-mode optical fibers with arbitrary refractive-index profile, and will be discussed in Section 4.6.

When the propagation constant becomes smaller than $n_2 k$, the mode is a radiation mode and the solution of Eq. (4.2-15) is not the exponentially decaying function but the sinusoidal function. The field distribution is thus expressed by the following:

Even radiation modes,

$$E_y = \begin{cases} C_e \cos(\kappa x), & |x| \leq a, \\ \cos(\kappa a)[C_e \cos\{\sigma(|x| - a)\} + C'_e \sin\{\sigma(|x| - a)\}], & x > a, \end{cases}$$
$$(4.2\text{-}26)$$

Odd radiation modes:

$$E_y = \begin{cases} C_0 \sin(\kappa x), & |x| \leq a, \\ (|x|/x) \sin(\kappa a)[C_0 \cos\{\sigma(|x| - a)\} + C'_0 \sin\{\sigma(|x| - a)\}] & |x| > a, \end{cases}$$
$$(4.2\text{-}27)$$

where

$$\sigma = \sqrt{n_2^2 k^2 - \beta^2}. \tag{4.2-28}$$

The propagation constant of a radiation mode belongs to a continuous set of eigenvalues in the range of $|\beta| < n_2 k$, because the existence of additional arbitrary constants C'_e and C'_0 makes the number of unknown variables exceed the number of equations derived from the boundary conditions. When the waveguide includes some imperfections, such as small irregularities of the core–cladding boundary and bending with a small radius, the mode conversion occurs between higher- and lower-order guided modes and also between guided and radiation modes. This mode conversion from guided to radiation modes causes the bending and scattering losses.

B. TM MODES

Starting from $H_z = 0$, the H_y component of TM modes is obtained in the same way.

Even modes $\quad H_y = \begin{cases} B_e \cos \kappa x, & |x| \leq a, \\ B_e \cos(\kappa a) \exp[-\gamma(|x| - a)] & |x| > a; \end{cases} \tag{4.2-29}$

Odd modes $\quad H_y = \begin{cases} B_0 \sin \kappa x & |x| \leq a, \\ (x/|x|) B_0 \sin(\kappa a) \exp[-\gamma(|x| - a)] & |x| > a. \end{cases} \tag{4.2-30}$

Another tangential field component E_z is obtained from H_y with the aid of Eq. (4.2-2):

$$E_z = \frac{1}{j\omega\varepsilon} \frac{\partial H_y}{\partial x}. \tag{4.2-31}$$

Since E_z must be continuous at the core–cladding boundary, the eigenvalue equations for TM even and odd modes are derived as shown in Table 4.2-I with the aid of Eqs. (4.2-29) through (4.2-31). The solutions of Eqs. (4.2-32) and (4.2-33) are also normalized by the parameters b and V, given by Eqs. (4.2-22) and (4.2-23). These solutions for TM modes differ from those for TE modes due to the $(n_1/n_2)^2$ and $(n_2/n_1)^2$ terms in Eqs. (4.2-32) and (4.2-33). The difference, however, is usually negligible, because n_1/n_2 can be approximated by unity in most cases. The cutoff V values of TM modes are exactly equal to those of TE modes when the mode number is equal.

C. MODE CONFINEMENT FACTOR

The mode confinement factor ξ is defined as the ratio of the optical power confined in the core region to the total power. This factor is important for semiconductor lasers having optical gain in the core region, because it is related to the mode gain. The mode confinement factor of the even TE mode is calculated from Eq. (4.2-16) as

$$\xi = \frac{\displaystyle\int_0^{d/2} |E_y|^2 \, dx}{\displaystyle\int_0^{\infty} |E_y|^2 \, dx} = \frac{1 + (\sin \kappa d/\kappa d)}{1 + (\sin \kappa d/\kappa d) + [2\cos^2(\kappa d/2)/\gamma d]}, \quad (4.2\text{-}34)$$

where $d = 2a$ according to the usual notation for semiconductor lasers. However, since Eq. (4.2-34) is not normalized by the index difference and the waveguide thickness, it is not as convenient for practical use. By using Eq. (4.2-20) we can express simply the confinement factor in terms of V and b as

$$\xi = \frac{V + \sqrt{b}}{V + 1/\sqrt{b}}. \quad (4.2\text{-}34')$$

Since the relation between V and b is given in Fig. 4.2-2, the confinement factor can be expressed as a function of V. The confinement factor ξ versus V is shown in Fig. 4.2-3 with other parameters κa, γa, b. For convenience, the confinement factor of the TE_0 mode versus the active layer thickness of GaInAsP–InP lasers is shown in Fig. 4.2-4.

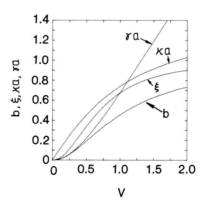

FIGURE 4.2-3. Confinement factor ξ and other parameters b, κa, and γa vs V.

In the case of GaInAsP–InP semiconductor lasers, the refractive index of the InP cladding layer is constant. On the other hand, the lasing wavelength and refractive index of the $Ga_x In_{1-x} As_y P_{1-y}$ active layer varies according to the compositions x and y. Therefore, the wavelength and refractive index difference change simultaneously in Fig. 4.2-4.

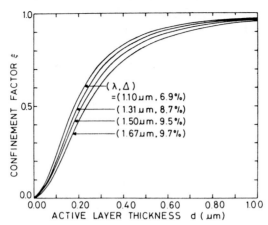

FIGURE 4.2-4.   The mode confinement factor ξ. (After Itaya, Suematsu, Katayama, Kishino, and Arai.[2])

D. FAR-FIELD PATTERN

We shall now consider the far-field pattern of modes. By doing this, we can find the relation of ray optics to the wave theory of slab waveguides.

Let us assume the far-field pattern observed in a plane far from the end of the waveguide to be L, as shown in Fig. 4.2-5. The far-field pattern can be calculated by the Fraunhofer diffraction and is expressed in terms of the near field by

$$E_y(x', L) \cong \frac{j}{\lambda L} \int_{-\infty}^{\infty} E_y(x, 0) \exp(-jk_0 r) \, dx, \qquad (4.2\text{-}35)$$

where $r^2 = (x' - x)^2 + L^2$. Equation (4.2-35) is the integral transform of $E_y(x, 0)$.

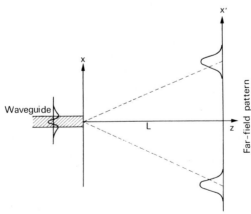

FIGURE 4.2-5.   Far-field pattern of a step-index planar waveguide.

Now we suppose an even TE mode, whose mode function is expressed by Eq. (4.2-16). Substitution of Eq. (4.2-16) into Eq. (4.2-35) gives

$$E_y(x', L) = \frac{j}{\lambda L} \left[\int_{-a}^{a} A_e \cos(\kappa x) \exp(-jk_0 r)\, dx \right.$$

$$+ \int_{a}^{\infty} A_e \cos(\kappa a) \exp[-\gamma(|x| - a)] \exp(-jk_0 r)\, dx$$

$$\left. + \int_{-\infty}^{-a} A_e \cos(\kappa a) \exp[-\gamma(|x| - a)] \exp(-jk_0 r)\, dx \right]. \quad (4.2\text{-}36)$$

The second and third terms on the right-hand side of Eq. (4.2-32) can be neglected, because the $\exp(-jk_0 r)$ term is a sinusoidal periodic function with a very short period compared with the decay of field function. Here, assuming $|x'|, L \gg a$, distance r can be approximated by

$$r \cong \sqrt{L^2 + x'^2} \left(1 - \frac{x'x}{L^2 + x'^2} \right). \quad (4.2\text{-}37)$$

By this approximation, Eq. (4.2-35) can be modified into a Fourier transform. By substituting Eq. (4.2-37) into Eq. (4.2-36) and doing some calculation, we obtain

$$E_y(x', L) = \frac{j}{\lambda L} A_e \exp(-jk_0 \sqrt{x'^2 + L^2})$$

$$\times \left[\frac{\sin[\kappa - (k_0 x'/\sqrt{x'^2 + L^2})]a}{\kappa - (k_0 x'/\sqrt{x'^2 + L^2})} + \frac{\sin[\kappa + (k_0 x'/\sqrt{x'^2 + L^2})]a}{\kappa + (k_0 x'/\sqrt{x'^2 + L^2})} \right].$$

$$(4.2\text{-}38)$$

Equation (4.2-34) shows that the main lobe of the far-field pattern is formed in the direction determined by

$$x'/\sqrt{x'^2 + L^2} = \pm(\kappa/k_0). \quad (4.2\text{-}39)$$

This direction coincides with the radiated direction of the plane waves that construct the mode field Eq. (4.2-16). This can be seen easily by resolving $\cos \kappa x$ in Eq. (4.2-16) into two plane waves as $\frac{1}{2}[\exp(j\kappa x) + \exp(-j\kappa x)]$ and considering the refraction at the end surface, as shown in Fig. 4.2-6. When $L, |x| \gg a$, Eq. (4.2-39) is rewritten as

$$x'/\sqrt{x'^2 + L^2} \cong \sin \theta_0 = n_1 \sin \theta_i'$$

$$= n_1(\kappa/k_0 n_1) = \kappa/k_0. \quad (4.2\text{-}40)$$

The radiation angle of the highest-order mode calculated from Eq. (4.2-40) gives the maximum radiation angle of the waveguide. The highest-order mode

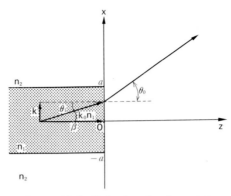

FIGURE 4.2-6.   Refraction of a plane wave.

can be regarded as a mode just at cutoff. Substituting Eq. (4.2-18) into Eq. (4.2-40) and putting $\beta = n_2 k$, we obtain,

$$\theta_0^{max} = \sin^{-1} \sqrt{n_1^2 - n_2^2}. \qquad (4.2-41)$$

Thus the maximum radiation angle from a multimode waveguide is determined by only the refractive index difference between the core and the cladding. The maximum acceptance angle is also determined by Eq. (4.2-41). Now we define a numerical aperture (NA) as

$$NA = n_1 \sin \theta_i^{max} = \sin \theta_0^{max} = \sqrt{n_1^2 - n_2^2} = n_1 \sqrt{2\Delta}. \qquad (4.2-42)$$

This is an important parameter concerning the maximum acceptance angle of incident light and the maximum radiation angle of output light.

On the other hand, readers may be questioning how the far-field pattern of the fundamental (lowest-order) mode is expressed. Since the near-field pattern of the fundamental mode has a single peak, it is expected that the far-field pattern will also have a single peak. This prospect is true and can be demonstrated as follows.

Two lobes of the far-field pattern expressed by Eq. (4.2-38) have the same form $\sin(ua)/u$, where $u = \kappa \pm k_0 x'/\sqrt{x'^2 + L^2}$. The angle θ_d between the first zero and the optical axis is derived from Eq. (4.2-40) as

$$\theta_d = \sin^{-1}[(\pi - \kappa a)/k_0 a]. \qquad (4.2-43)$$

On the other hand, the angle between the main peak and the optical axis is expressed from Eq. (4.2-40) as

$$\theta_m = \sin^{-1}(\kappa a/k_0 a). \qquad (4.2-44)$$

For the fundamental mode, κa must lie in the range of $0 < \kappa a < \pi/2$, because the field described by Eq. (4.2-16) has only one peak in the core. Therefore, the angles θ_d and θ_m also must lie, respectively, in the ranges of

$$\sin^{-1}\left(\frac{\pi}{2k_0 a}\right) < \theta_d < \sin^{-1}\left(\frac{\pi}{k_0 a}\right),$$

$$0 < \theta_m < \sin^{-1}\left(\frac{\pi}{2k_0 a}\right). \tag{4.2-45}$$

Thus the far-field pattern, which is expressed by the superposition of these two main lobes, also has a single peak as shown in Fig. 4.2-7, because the diffraction angle θ_d of one main lobe is larger than the separation angle of two main lobes.

By referring to these discussions, we note that the numerical aperture of a single mode waveguide is not purely described as a function of $\sqrt{2\Delta}$ but must include the wavelength and the width of waveguide.

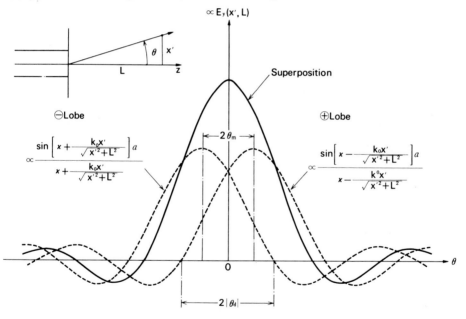

FIGURE 4.2-7.   Far-field pattern of a fundamental mode.

4.3 DISTRIBUTED-INDEX PLANAR WAVEGUIDE

In this section we suppose a distributed index planar waveguide whose refractive index distribution varies in proportion to the square of transverse distance x. The study of this type of planar waveguides is important for understanding eigenmodes of distributed-index multimode fibers, rod lenses, integrated optical waveguides, and semiconductor lasers that have this type of waveguide.

We express the refractive index distribution as

$$n^2(x) = \begin{cases} n_1^2[1 - (gx)^2], & |x| \le a, \\ n_2^2, & |x| > a, \end{cases} \tag{4.3-1}$$

where n_1 and n_2 are refractive indices of core center and cladding, respectively and $g = \sqrt{2\Delta}/a$ and is called a focusing constant. As mentioned in Section 4.2, wave equations (4.2-6) and (4.2-7) for distributed-index waveguides include gradient terms of ε. The gradient term of Eq. (4.2-6) can be evaluated as

$$\left| \mathbf{V}\left(\frac{\mathbf{V}\varepsilon}{\varepsilon} \cdot \mathbf{E} \right) \right| \simeq g^2 E_t, \tag{4.3-2}$$

where E_t is the transverse component of electric field. On the other hand, the time derivative part of wave equation (4.2-6) is evaluated as

$$\omega^2 \mu_0 \varepsilon E_t \simeq k_0^2 n^2 E_t. \tag{4.3-3}$$

The ratio of Eq. (4.3-2) to Eq. (4.3-3) is on the order of g^2/k_0^2 in magnitude. When the width (or thickness) of a waveguide is large, this ratio becomes $\sim 10^{-6}$ to 10^{-7} and the gradient terms can be ignored. This is called a "weakly guiding approximation." In this way, we obtain a scalar wave equation

$$\mathbf{V}^2\mathbf{E} + k_0^2 n_1^2[1 - (gx)^2]\mathbf{E} = 0. \tag{4.3-4}$$

For TE modes, assuming z dependence to be $e^{-j\beta z}$, Eq. (4.3–4) becomes

$$H_0 E_y = \beta^2 E_y, \tag{4.3-5}$$

where

$$H_0 = \frac{\partial^2}{\partial x^2} + k_0^2 n_1^2(1 - g^2 x^2) \tag{4.3-6}$$

Equation (4.3-5) is an eigenvalue equation for an operator H_0 and β^2 is the eigenvalue. Here, introducing a new parameter w_0 and a new variable $\bar{x}$ defined by

$$w_0 = 1/\sqrt{k_0 n_1 g} \tag{4.3-7}$$

$$\bar{x} = x/w_0, \tag{4.3-8}$$

we transform Eq. (4.3-5) into

$$\frac{d^2 E_y}{d\bar{x}^2} + [\kappa^2 w_0^1 - \bar{x}^2] E_y = 0, \tag{4.3-9}$$

where

$$\kappa^2 = k_0^2 n_1^2 - \beta^2. \tag{4.3-10}$$

The solution of Eq. (4.3-9), which reaches zero at $\bar{x} = \pm\infty$, is given[3] in terms of the Hermite polynomial $H_p(\bar{x})$ by

$$E_y(\bar{x}) = N_p H_p(\bar{x}) \exp(-\tfrac{1}{2}\bar{x}^2). \tag{4.3-11}$$

Here, the Hermite polynomials of the pth order are tabulated in Table 4.3-I.

TABLE 4.3-I

HERMITE POLYNOMIALS.

p	$H_p(\xi)$
0	1
1	2ξ
2	$4\xi^2 - 2$
3	$8\xi^3 - 12\xi$
$\vdots$	$\vdots$

Since the eigenvalue of Eq. (4.3-9) is given by

$$\kappa^2 w_0^2 = 2p + 1, \tag{4.3-12}$$

the propagation constant β is

$$\begin{aligned}
\beta &= k_0 n_1 [1 - (g/k_0 n_1)(2p + 1)]^{1/2} \\
&\simeq k_0 n_1 - g(p + \tfrac{1}{2}).
\end{aligned} \tag{4.3-13}$$

The propagation constant β of guided modes must be in the range of $n_1 k_0 > \beta > n_2 k_0$. Therefore, the total number of guided modes can be approximately evaluated as

$$p_{max} \simeq \tfrac{1}{2} k_0 n_1 a \sqrt{2\Delta} = \tfrac{1}{2} V. \tag{4.3-14}$$

On the other hand, from the normalization condition

$$\int_{-\infty}^{\infty} E_y(x)\, dx = 1, \tag{4.3-15}$$

the normalization constant N_p in Eq. (4.3-11) is calculated as

$$N_p = 1/[2^p p! \sqrt{\pi} w_0]^{1/2}. \tag{4.3-16}$$

Thus, we obtain the eigenfunction

$$E_y(x) = \frac{1}{[2^p p! \sqrt{\pi} w_0]^{1/2}} H_p\left(\frac{x}{w_0}\right) \exp\left[-\frac{1}{2}\left(\frac{x}{w_0}\right)^2\right]. \tag{4.3-17}$$

The mode function expressed by Eq. (4.3-17) is called a Hermite–Gaussian mode function, because it is the product of the Hermite polynomials and Gaussian function. The parameter w_0 defined by Eq. (4.3-7) is called a characteristic spot size. Figure 4.3-1 shows the mode function Eq. (4.3-17) for $p = 0, 1, 2$, and 3. It is easily seen from the figure that the mode number p corresponds to the number of zeros in the field distribution.

The Hermite–Gaussian mode is an approximation, because in Eq. (4.3-4) the index distribution is supposed to reach $-\infty$ at $x \to \infty$. The actual distribution takes a constant value n_2 in the cladding. However, when the mode field is well confined in the core region, the Hermite–Gaussian mode provide a good approximation.

As for the far-field pattern of a parabolic DI waveguide, we can calculate it as follows. As mentioned after Eq. (4.2-37), the far-field pattern can be expressed by the Fourier transform of the near field. Since the Fourier transform of Eq. (4.3-9) has also the same form, we obtain

$$E_y(x', L) = \frac{j^{p+1} \sqrt{2\pi} \, N_p^{\omega_0}}{\lambda L} \exp(-jk_0\sqrt{x'^2 + L^2})H_p(\Phi) \exp(-\tfrac{1}{2}\Phi^2) \quad (4.3\text{-}18)$$

where

$$\Phi = k_0 w_0 x' / \sqrt{x'^2 + L^2} = k_0 w_0 \sin\theta. \quad (4.3\text{-}19)$$

This equation implies that the far-field pattern of a parabolic DI waveguide is only a function of angle θ measured from the optical axis. Thus the field spreads in proportion to the distance but is not altered in its shape.

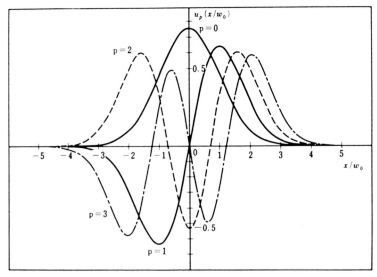

FIGURE 4.3-1.   Hermite–Gaussian mode functions $u_p(x/w_0)$.

The maximum radiation angle of parabolic DI waveguide can be calculated as follows: Eq. (4.3-18) is similar to the solution of a harmonic oscillator in quantum mechanics and the corresponding amplitude Φ_0 of a classical harmonic oscillator can be expressed in terms of mode number p by[3]

$$\Phi_0 = \sqrt{2p}, \tag{4.3-20}$$

where Φ_0 can be regarded as the width of pth-order Hermite–Gaussian mode. On the other hand, the maximum mode number p_{max} can be calculated by putting $\beta = k_0 n_2$ in Eq. (4.3-13) as

$$p_{max} = \frac{1}{g} k_0 (n_1 - n_2) - \frac{1}{2} \simeq \frac{k_0 n_1 \Delta}{g} \tag{4.3-21}$$

By using Eqs. (4.3-19)–(4.3-21) and Eq. (4.3-7) we obtain the NA of a parabolic DI waveguide as

$$NA = \sin \theta_0^{max} = \frac{1}{k_0 w_0} \sqrt{2 p_{max}} = n_1 \sqrt{2\Delta}. \tag{4.3-22}$$

It should be noted here that the NA of a parabolic DI waveguide is the same as that of a step index waveguide and can be determined only by the index difference.

4.4 STEP-INDEX ROUND FIBER AND ROD

In this section we deal with a round optical fiber whose refractive index distribution is constant in the core and in the cladding (see Fig. 4.4-1). We express the refractive index distribution as

$$n^2(r) = \begin{cases} n_1^2 & \text{for } r \leq a, \\ n_2^2 & \text{for } r > a. \end{cases} \tag{4.4-1}$$

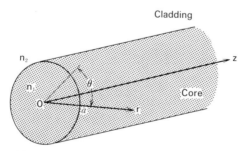

FIGURE 4.4-1.   A step-index round optical fiber.

This type of round optical fiber is called the step-index fiber. The wave equations for the step-index fiber are also given by Eqs. (4.2-6) and (4.2-7). The gradient-index terms of these equations, however, vanish because of the fiber's constant refractive index in the core and cladding. Now we shall deal with field components in terms of cylindrical coordinates. The wave equations for E_z and H_z are given by

$$\frac{\partial^2 E_z}{\partial r^2} + \frac{1}{r}\frac{\partial E_z}{\partial r} + \frac{1}{r^2}\frac{\partial^2 E_z}{\partial \theta^2} + (k_0^2 n^2 - \beta^2)E_z = 0, \tag{4.4-2}$$

$$\frac{\partial^2 H_z}{\partial r^2} + \frac{1}{r}\frac{\partial H_z}{\partial r} + \frac{1}{r^2}\frac{\partial^2 H_z}{\partial \theta^2} + (k_0^2 n^2 - \beta^2)H_z = 0, \tag{4.4-3}$$

where $k_0^2 = \omega^2 \mu_0 \varepsilon_0$. The transverse field components are expressed in terms of E_z and H_z by

$$E_r = -j\frac{\beta}{\kappa^2}\left(\frac{\partial E_z}{\partial r} + \frac{\omega\mu}{\beta}\frac{1}{r}\frac{\partial H_z}{\partial \theta}\right), \tag{4.4-4}$$

$$E_\theta = -j\frac{\beta}{\kappa^2}\left(\frac{1}{r}\frac{\partial E_z}{\partial \theta} - \frac{\omega\mu}{\beta}\frac{\partial H_z}{\partial r}\right), \tag{4.4-5}$$

$$H_r = -j\frac{\beta}{\kappa^2}\left(-\frac{k_0^2 n^2}{\omega\mu\beta}\frac{1}{r}\frac{\partial E_z}{\partial \theta} + \frac{\partial H_z}{\partial r}\right), \tag{4.4-6}$$

$$H_\theta = -\frac{1}{j}\frac{\beta}{\kappa^2}\left(\frac{k_0^2 n^2}{\omega\mu\beta}\frac{\partial E_z}{\partial r} + \frac{1}{r}\frac{\partial H_z}{\partial \theta}\right). \tag{4.4-7}$$

It should be noted that the wave equations (4.4-2) and (4.4-3) are independent. However, the guided modes in a round optical fiber cannot, except for one special case, be classified simply into TE and TM modes as with planar waveguides. This is because the transverse field components E_θ and H_θ include both E_z and H_z. The boundary condition is that the tangential field components E_θ, H_θ, E_z and H_z are continuous at the core–cladding boundary. If either E_z or H_z is zero, the number of equations exceeds the number of variables; thus, all of the boundary conditions cannot always be satisfied simultaneously. Therefore, the guided modes generally include all six field components. These modes are called hybrid modes.

A. CHARACTERISTIC EQUATION

The solution of Eqs. (4.4-2) and (4.4-3) is given by

$$E_z = A_l J_l(\kappa r)\cos(l\theta + \varphi_l)\exp[j(\omega t - \beta z)] \tag{4.4-8}$$

$$H_z = B_l J_l(\kappa r)\cos(l\theta + \psi_l)\exp[j(\omega t - \beta z)] \tag{4.4-9}$$

in the core ($|r| \leq a$) and

$$E_z = C_l K_l(\gamma r) \cos(l\theta + \varphi_l) \exp[j(\omega t - \beta z)] \tag{4.4-10}$$

$$H_z = D_l K_l(\gamma r) \cos(l\theta + \psi_l) \exp[j(\omega t - \beta z)] \tag{4.4-11}$$

in the cladding ($r > a$). Here J_l and K_l are the Bessel function and modified Hankel function of the first kind and κ and γ are defined by Eqs. (4.2-18) and (4.2-19). Tangential components E_θ and H_θ are obtained by substituting Eqs. (4.4-8)–(4.4-11) into Eqs. (4.4-5) and (4.4-7). From the boundary condition stated above, a characteristic equation is derived:[4,5]

$$k^2(\eta_1 + \eta_2)(n_1^2\eta_1 + n_2^2\eta_2) = l^2\beta^2 \left(\frac{1}{u^2} + \frac{1}{w^2}\right)^2, \tag{4.4-12}$$

where

$$\eta_1 = J_l'(u)/uJ_l(u), \tag{4.4-13}$$

$$\eta_2 = K_l'(w)/wK_l(w), \tag{4.4-14}$$

$$u = \sqrt{k_0^2 n_1^2 - \beta^2}\,a, \tag{4.4-15}$$

$$w = \sqrt{\beta^2 - k_0^2 n_2^2}\,a = \sqrt{V^2 - u^2}, \tag{4.4-16}$$

and the phase angles φ and ψ must satisfy

$$\varphi - \psi = \pm\pi/2. \tag{4.4-17}$$

The propagation constant β can be obtained by solving the above transcendental equation.

B. LP MODES

Equation (4.4-12) is rather complicated and not very useful, except for the $l = 0$ case. When $l = 0$, Eq. (4.4-12) separates into two independent equations:

$$\eta_1 + \eta_2 = 0, \tag{4.4-18}$$

$$n_1^2\eta_1 + n_2^2\eta_2 = 0. \tag{4.4-19}$$

Equation (4.4-18) corresponds to TE modes and Eq. (4.4-19) to TM modes. The reason can be easily seen, since the substitution of Eqs. (4.4-18) and (4.4-19) into the boundary conditions gives $E_z = 0$ and $H_z = 0$, respectively.

However, in the general case that $l \neq 0$, the guided modes are hybrid modes. To simplify the problem, we introduce a new parameter P defined by

$$P \equiv -\frac{\omega\mu}{\beta} \frac{B_l \cos(l\theta + \psi_l)}{A_l \sin(l\theta + \phi_l)}, \tag{4.4-20}$$

which indicates the relative ratio of E_z and H_z. The field components in the core are expressed in terms of P as

$$E_z = J_l(\kappa r) F_c, \tag{4.4-21}$$

$$E_r = -j\frac{\beta}{\kappa}\left[J_l'(\kappa r) - P\frac{l}{\kappa r} J_l(\kappa r) \right] F_c \tag{4.4-22}$$

$$E_\theta = -j\frac{\beta}{\kappa}\left[PJ_l'(\kappa r) - \frac{l}{\kappa r} J_l(\kappa r) \right] F_s, \tag{4.4-23}$$

$$H_z = -\frac{\beta}{\omega\mu} PJ_l(\kappa r) F_s, \tag{4.4-24}$$

$$H_r = -j\frac{k_0^2 n_1^2}{\omega\mu\kappa}\left[P\left(\frac{\beta^2}{k_0^2 n_1^2}\right) J_l'(\kappa r) - \frac{l}{\kappa r} J_l(\kappa r) \right] F_s, \tag{4.4-25}$$

$$H_\theta = -j\frac{k_0^2 n_1^2}{\omega\mu\kappa}\left[J_l'(\kappa r) - P\frac{\beta^2}{k_0^2 n_1^2}\frac{l}{\kappa r} J_l(\kappa r) \right] F_c, \tag{4.4-26}$$

where the prime denotes the differentiation of the Bessel function with respect to the argument, and F_c and F_s are given by

$$F_c = A_l \cos(l\theta + \varphi_l) \exp[j(\omega t - \beta z)], \tag{4.4-27}$$

$$F_s = A_l \sin(l\theta + \varphi_l) \exp[j(\omega t - \beta z)], \tag{4.4-28}$$

Now we shall apply the "weakly guiding approximation,"[6] where the index difference is much smaller than unity ($\Delta \ll 1$). Since $k_0 n_1 > \beta > k_0 n_2$ for guided modes, this approximation used in Eq. (4.4-12) yields

$$\eta_1 + \eta_2 = \pm l\left(\frac{1}{u^2} + \frac{1}{w^2}\right). \tag{4.4-29}$$

By using the formulas of Bessel and modified Hankel functions, Eq. (4.4-29) is rewritten as

$$J_{l+1}(u)/uJ_l(u) = -K_{l+1}(w)/wK_l(w) \tag{4.4-30}$$

for the positive values of Eq. (4.4-29) and

$$J_{l-1}(u)/uJ_l(u) = K_{l-1}(w)/wK_l(w) \tag{4.4-31}$$

for the negative values. These are the characteristic equations for the weakly guiding case.

By using the boundary conditions, the parameter P is expressed in terms of $\eta_1, \eta_2, u,$ and w as

$$P = l\left(\frac{1}{u^2} + \frac{1}{w^2}\right)\bigg/(\eta_1 + \eta_2) \tag{4.4-32}$$

Substitution of Eq. (4.4-29) into Eq. (4.4-32) gives

$$P = \pm 1 \qquad (4.4\text{-}33)$$

The positive case is called the EH mode and the negative, the HE mode. Hence Eq. (4.4-30) is the characteristic equation for EH modes and Eq. (4.4-31), for HE modes. However, by using the recurrence formulas for Bessel and modified Hankel functions, Eq. (4.4-31) can be modified as

$$J_{l-1}(u)/uJ_{l-2}(u) = -K_{l-1}(w)/wK_{l-2}(w). \qquad (4.4\text{-}34)$$

Equations (4.4-30) and (4.4-34) have the same form, and they can be written uniformly as

$$J_\mu(u)/uJ_{\mu-1}(u) = -K_\mu(w)/wK_{\mu-1}(w) \qquad (4.4\text{-}35)$$

where

$$\mu = \begin{cases} l+1 & \text{for EH modes,} \\ l-1 & \text{for HE modes.} \end{cases} \qquad (4.4\text{-}36)$$

This implies that the $EH_{\mu-1,m}$ mode and $HE_{\mu+1,m}$ mode are degenerated with the propagation constant.

This degenerated mode group is called the $LP_{\mu,m}$ mode, where LP is taken from linear polarized modes and means that the linear combination of degenerated modes $(EH_{\mu-1,m} \pm HE_{\mu+1,m})$ constructs a linear polarized transverse field.[6] Actually, this can be seen by using Eqs. (4.4-22)–(4.4-26), and (4.4-33) as follows. The transverse field E_x in the Cartesian coordinates is expressed by using Eqs. (4.4-22) and (4.4-23) as

$$E_x^l = -j\frac{\beta}{\kappa}\left[(F_c \cos\theta - PF_s \sin\theta)\, J_l'(\kappa r) - \frac{l}{\kappa r}(F_c P \cos\theta - F_s \sin\theta)\, J_l(\kappa r) \right]$$

$$(4.4\text{-}37)$$

The sum of $E_x^{(\mu-1)}$ for $EH_{\mu-1,m}$ mode $(P = 1)$ and $E_x^{(\mu+1)}$ for $HE_{\mu+1,m}$ mode $(p = -1)$ can be shown to be zero with formulas for the Bessel function. Figure 4.4-2 shows the dispersion curve of the $LP_{\mu,m}$ modes.

As for the LP_{11} mode, the process to produce a linear polarized field from the superposition of HE_{21}, TE_{01}, and TM_{01} modes is rather complicated. The E_x of TE_{01} and TM_{01} modes are derived from the substitution of Eqs. (4.4-8) and (4.4-9) into Eqs. (4.4-4)–(4.4-7) as follows:

$$E_x^{(TE_{01})} = E_r \cos\theta - E_\theta \sin\theta = -j\frac{\omega\mu}{\kappa} B_0 \cos\psi_0 \sin\theta J_1(\kappa r), \quad (4.4\text{-}38)$$

$$E_x^{(TM_{01})} = -j\frac{\beta}{\kappa} A_0 \cos\varphi_0 \cos\theta J_1(\kappa r). \qquad (4.4\text{-}39)$$

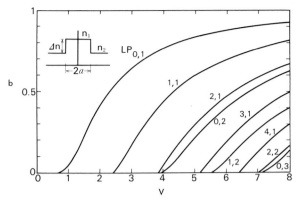

FIGURE 4.4-2. Dispersion curve of a step-index round fiber. (After Tanaka and Suematsu.[7])

On the other hand, the E_x of the HE_{21} mode is obtained from Eq. (4.4-37) as

$$E_x^{(HE_{21})} = -j\frac{\beta}{\kappa} A_2 \cos(\theta + \varphi_2)J_1(\kappa r). \qquad (4.4\text{-}40)$$

Thus we can obtain the following equations by choosing arbitrary constants $A_0, B_0, A_2, \varphi_0, \psi_0,$ and φ_2 in Eqs. (4.4-38)–(4.4-40). For example, by putting $\omega\mu B_0 = \beta A_2, \varphi_0 = 0$ and $\varphi_2 = \pi/2$, we obtain

$$E_x^{(TE_{01})} + E_x^{(HE_{21})} = 0, \qquad (4.4\text{-}41)$$

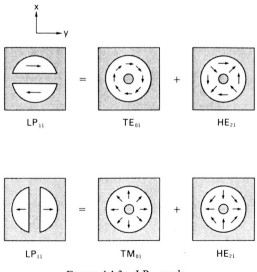

FIGURE 4.4-3. LP_{11} mode.

and by putting $A_0 = A_2$, $\varphi_0 = 0$, and $\varphi_2 = 0$,

$$E_x^{(\text{TM}_{01})} + E_x^{(\text{HE}_{21})} = 0. \tag{4.4-42}$$

Thus we can derive LP_{11} mode by the superposition of TE_{01}, TM_{01}, and HE_{21} modes as shown in Fig. 4.4-3.

As for the fundamental LP_{01} mode, it is easily seen from Eq. (4.4-37) that the field E_x can be zero by choosing the arbitrary phase $\varphi_1 = \pi/2$.

Since the degeneracy of LP modes results from the weakly guiding approximation, the exact values of the propagation constant β of the $\text{EH}_{\mu-1,m}$ mode and the $\text{HE}_{\mu+1,m}$ mode are not equal to each other. But since the difference is expected to be very small, the weakly guiding approximation is a useful one. Another useful but rather rough approximation is the principal-mode approximation. This will be used in the next section. The relations between exact and approximated modes are summarized in Table 4.4-I.

TABLE 4.4-I

RELATIONS BETWEEN EXACT AND APPROXIMATED MODES

Exact modes	LP mode approximation	Principal mode approximation
HE_{11}	LP_{01}	
TE_{0m} TM_{0m} HE_{2m}	LP_{1m}	$N = 2\mu + m$
$\text{EH}_{\mu-1,m}$ $\text{HE}_{\mu+1,m}$ $(\mu \geq 2)$	$\text{LP}_{\mu m}$	N, principal mode number

4.5 DISTRIBUTED-INDEX ROUND FIBER

The distributed-index round fiber is well known as the transmission media of optical communications and focusing-type rod lenses. This type of fiber has many useful features, such as a large bandwidth when used as a transmission media and focusing and imaging properties when used as a microlens. In this section, we shall derive the modes of the DI round fiber by using two different coordinate systems—the Cartesian and the cylindrical.

A. HERMITE–GAUSSIAN MODES

The index distribution in the Cartesian coordinate system is given as

$$n^2(x, y) = \begin{cases} n_1^2[1 - (g_1 x)^2 - (g_2 y)^2] & \text{for} \quad |x| \leq a, \quad |y| \leq b, \\ n_2^2 & \text{for} \quad |x| > a, \quad |y| > b, \end{cases} \tag{4.5-1}$$

where g_1 and g_2 are the focusing constants introduced in Eq. (4.3-1), and n_1 and n_2 the refractive indices of the core center and the cladding, respectively. Substituting Eq. (4.5-1) into Eq. (4.2-6) and neglecting the gradient-index term, we obtain the following scalar wave equation, similar to Eq. (4.3-4):

$$\mathbf{V}^2\mathbf{E} + k_0^2 n_1^2[1 - (g_1 x)^2 - (g_2 y)^2]\mathbf{E} = 0. \tag{4.5-2}$$

Here we assume the x polarized field. This approximation corresponds to LP modes, but it is usually called the TEM wave (transverse electromagnetic wave) approximation. Since the z component can be obtained from Maxwell's equations, we work on the E_x component of Eq. (4.5-2):

$$\mathbf{V}E_x + k_0^2 n_1^2[1 - (g_1 x)^2 - (g_2 y)^2]E_x = 0 \tag{4.5-3}$$

Assuming the z dependence as $\exp(-j\beta z)$, we express the E_x component in terms of the separated variables as

$$E_x = X(x)Y(y)\exp(-j\beta z). \tag{4.5-4}$$

Substitution of Eq. (4.5-4) into Eq. (4.5-3) gives

$$\frac{1}{X}\frac{d^2 X}{dx^2} - k_0^2 n_1^2(g_1 x)^2 + \frac{1}{Y}\frac{d^2 Y}{dy^2} - k_0^2 n_1^2(g_2 y)^2 = \beta^2 - k_0^2 n_1^2. \tag{4.5-5}$$

With the condition that both sides of Eq. (4.5-5) must be constant with respect to x and y, the following equations are obtained.

$$\frac{1}{X}\frac{d^2 X}{dx^2} - k_0^2 n_1^2 g_1^2 x^2 = -\kappa_1^2, \tag{4.5-6}$$

$$\frac{1}{Y}\frac{d^2 Y}{dy^2} - k_0^2 n_1^2 g_2^2 y^2 = -\kappa_2^2, \tag{4.5-7}$$

$$\beta^2 = k_0^2 n_1^2 - \kappa_1^2 - \kappa_2^2. \tag{4.5-8}$$

Equation (4.5-6) is then rewritten as

$$\frac{d^2 X}{dx^2} + (\kappa_1^2 - k_0^2 n_1^2 g_1^2 x^2)X = 0. \tag{4.5-9}$$

Furthermore, by defining the characteristic spot size by

$$w_{01} = 1/\sqrt{k_0 n_1 g_1} \tag{4.5-10}$$

and transforming the variable as

$$\bar{x} = x/w_{01}, \tag{4.5-11}$$

Eq. (4.5-9) becomes

$$\frac{d^2 X}{d\bar{x}^2} + (\kappa_1^2 w_{01}^2 - \bar{x}^2)X = 0. \tag{4.5-12}$$

This equation is the same as Eq. (4.3-9), and the solution is given by

$$X(\bar{x}) = N_p H_p(\bar{x}) \exp(-\tfrac{1}{2}\bar{x}^2), \qquad (4.5\text{-}13)$$

where N_p is the normalized constant expressed by Eq. (4.3-16). The eigenvalue is given by

$$\kappa_1^2 = k_0 n_1 g_1 (2p + 1). \qquad (4.5\text{-}14)$$

On the other hand, the solution of $Y(y)$ can be obtained in the same manner and is given by

$$Y(\bar{y}) = N_q H_q(\bar{y}) \exp(-\tfrac{1}{2}\bar{y}^2), \qquad (4.5\text{-}15)$$

where q is the mode number of y direction and $\bar{y}$ is defined by

$$\bar{y} = y/w_{02} \qquad (4.5\text{-}16)$$

with

$$w_{02} = 1/\sqrt{k_0 n_1 g_2}. \qquad (4.5\text{-}17)$$

The eigenvalue is given by

$$\kappa_2^2 = k_0 n_1 g_2 (2q + 1). \qquad (4.5\text{-}18)$$

Thus, by substituting Eqs. (4.5-13) and (4.5-15) into Eq. (4.5-4), the mode function may be expressed as

$$E_x(x, y) = \frac{1}{[2^{p+q} p! q! \pi w_{01} w_{02}]^{1/2}} H_p\left(\frac{x}{w_{01}}\right) H_q\left(\frac{y}{w_{02}}\right)$$

$$\times \exp\left\{-\frac{1}{2}\left[\left(\frac{x}{w_{01}}\right)^2 + \left(\frac{y}{w_{02}}\right)^2\right]\right\} \exp(-j\beta z). \quad (4.5\text{-}19)$$

The propagation constant is also obtained by substituting Eqs. (4.5-14) and (4.5-18) into Eq. (4.5-8) as

$$\beta = \sqrt{k_0^2 n_1^2 - k_0 n_1 g_1 (2p + 1) - k_0 n_1 g_2 (2q + 1)}. \qquad (4.5\text{-}20)$$

In most cases in which the core radius is much larger than the wavelength, the following approximation holds:

$$k_0 \gg g_1(2p + 1), g_2(2q + 1). \qquad (4.5\text{-}21)$$

Equation (4.5-20) can then be approximated by

$$\beta_{pq} \simeq k_0 n_1 - (p + \tfrac{1}{2})g_1 - (q + \tfrac{1}{2})g_2. \qquad (4.5\text{-}22)$$

With DI round fibers, i.e., $g_1 = g_2 = g$, Eq. (4.5-22) becomes

$$\beta_{pq} \simeq k_0 n_1 - (p + q + 1)g. \qquad (4.5\text{-}23)$$

Equation (4.5-23) shows that a group of Hermite–Gaussian modes HG_{pq} for which $p + q$ equals a constant integer becomes degenerate. It will be shown later that this integer corresponds to the principal mode number N.

The group velocity v_g of HG_{pq} mode can be obtained by differentiating Eq. (4.5-23) with respect to k_0 with the velocity of light c as

$$v_g = c/(d\beta/dk_0) = c/n_1. \tag{4.5-24}$$

This implies that the group-delay difference of modes in DI round fibers is almost zero and that the DI round fiber has a very large bandwidth as a transmission media.

B. LAGUERRE GAUSSIAN MODES

The same index distribution is given in the cylindrical coordinate system as

$$n^2(r) = \begin{cases} n_1^2[1 - (gr)^2], & r \leq a, \\ n_2^2, & r > a. \end{cases} \tag{4.5-25}$$

Under the TEM wave approximation, the scalar wave equation for E_x in the cylindrical coordinate system can be derived from Eq. (4.2-6) by neglecting the gradient index terms:

$$\frac{\partial^2 E_x}{\partial r^2} + \frac{1}{r}\frac{\partial E_x}{\partial r} + \frac{1}{r^2}\frac{\partial^2 E_x}{\partial \theta^2} + (k_0^2 n^2(r) - \beta^2)E_x = 0, \tag{4.5-26}$$

where the z dependence $\exp(-j\beta z)$ was assumed. Substitution of Eq. (4.5-25) into Eq. (4.5-26) gives

$$\frac{\partial^2 E_x}{\partial r^2} + \frac{1}{r}\frac{\partial E_x}{\partial r} + \frac{1}{r^2}\frac{\partial^2 E_x}{\partial \theta^2} + [k_0^2 n_1^2(1 - g^2 r^2) - \beta^2]E_x = 0. \tag{4.5-27}$$

After the separation of variables,

$$E_x = R(r)\Theta(\theta), \tag{4.5-28}$$

Equation (4.5-27) is separated into the following three equations:

$$\frac{d^2 R}{dr^2} + \frac{1}{r}\frac{dR}{dr} - \left[k_0^2 n_1^2 g^2 r^2 + C_1 + \frac{C_2}{r^2}\right] R = 0, \tag{4.5-29}$$

$$d^2\Theta/d\theta^2 + C_2\Theta = 0, \tag{4.5-30}$$

$$\beta^2 = k_0^2 n_1^2 + C_1. \tag{4.5-31}$$

The solution of Eq. (4.5-30) is given by

$$\Theta(\theta) = \cos(l\theta + \zeta_l), \tag{4.5-32}$$

where

$$C_2 = l^2. \tag{4.5-33}$$

By using Eq. (4.5-33), Eq. (4.5-29) is rewritten as

$$\frac{d^2R}{d\bar{r}^2} + \frac{1}{r}\frac{dR}{d\bar{r}} - \left[\bar{r}^2 + C_1 w_0^2 + \frac{l^2}{\bar{r}^2}\right]R = 0. \tag{4.5-34}$$

Here, we have transformed the variable as

$$\bar{r} = r/w_0, \tag{4.5-35}$$

where w_0 is defined by

$$w_0 = 1/\sqrt{k_0 n_1 g}. \tag{4.5-36}$$

The solution of Eq. (4.5-34) is given by

$$R = (r/w_0)^m L_l^m(r^2/w_0^2) \exp(-\tfrac{1}{2}(r/w_0)^2), \tag{4.5-37}$$

where $L_l^m(x)$ is the associated Laguerre polynomial, defined by

$$L_l^m(x) = x^{-m} e^x \frac{d^l}{dx^l}(x^{l+m} e^{-x}). \tag{4.5-38}$$

The first several orders of associated Laguerre polynomials are listed in Table 4.5-I. The mode function is expressed by

$$E_x = N_{lm} \cos(l\theta + \zeta_l)(r/w_0)^m L_l^m(r^2/w_0^2) \exp[-\tfrac{1}{2}(r/w_0)^2 \exp(-j\beta z), \tag{4.5-39}$$

where N_{lm} is the normalization constant and is calculated as

$$N_{lm} = \begin{cases} \dfrac{\sqrt{\pi}}{\sqrt{(l!)^2 w_0}}, & m = 0 \\[2ex] \dfrac{\sqrt{2\pi}}{\sqrt{(l!)(l+m)! w_0}}, & m \geq 1 \end{cases} \tag{4.5-40}$$

TABLE 4.5-I

ASSOCIATED LAGUERRE POLYNOMIALS

	Laguerre polynomial $L_l^m(\xi)$
0	1
1	$1 - \xi$
2	$(m+2)(m+1) - 2(m+2)\xi + \xi^2$
3	$(m+3)(m+2)(m+1) - 3(m+3)(m+2)\xi$ $+ 3(m+3)\xi^2 - \xi^3$
$\vdots$	$\vdots$

The eigenvalue of Eq. (4.5-34) is given by

$$C_1 = -2(1/w_0^2)(m + 2l + 1) = -2k_0 n_1 g(m + 2l + 1). \qquad (4.5\text{-}41)$$

By substituting Eq. (4.5-41) into Eq. (4.5-31) and assuming

$$k_0 n_1 \gg g(m + 2l + 1),$$

the propagation constant is expressed by

$$\beta_{lm} \simeq k_0 n_1 - (m + 2l + 1)g. \qquad (4.5\text{-}42)$$

Equation (4.5-42) also shows that the group of Laguerre–Gaussian modes LG_{lm} for which $m + 2l$ equals a constant integer N becomes degenerate.

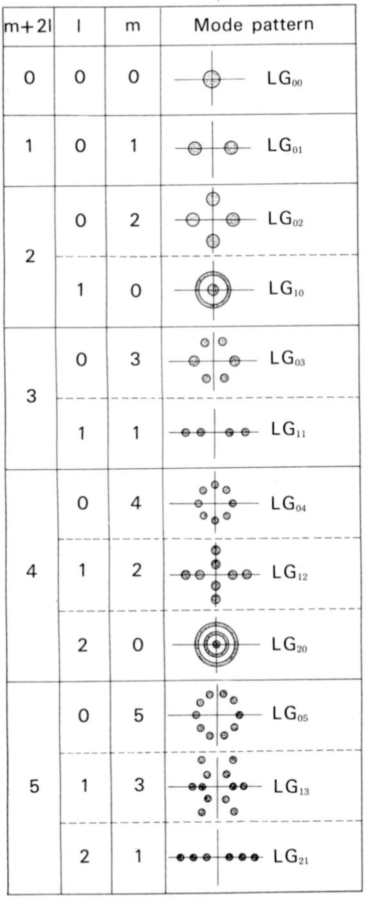

m+2l	l	m	Mode pattern
0	0	0	LG_{00}
1	0	1	LG_{01}
2	0	2	LG_{02}
	1	0	LG_{10}
3	0	3	LG_{03}
	1	1	LG_{11}
4	0	4	LG_{04}
	1	2	LG_{12}
	2	0	LG_{20}
5	0	5	LG_{05}
	1	3	LG_{13}
	2	1	LG_{21}

FIGURE 4.5-1. Mode pattern of Laguerre–Gaussian mode.

Therefore $N = m + 2l$ corresponds to the principal mode number. By comparing Eqs. (4.5-23) and (4.5-42), it is easily seen that the Hermite–Gaussian mode group HG_{pq} has the same propagation constant value as the Laguerre–Gaussian mode LG_{lm} when $p + q = m + 2l$. This means that a Laguerre–Gaussian mode can be expressed in terms of the linear combination of Hermite–Gaussian modes that have the same principal mode number.

Now we can calculate and illustrate the mode pattern by squaring Eq. (4.5-39). Figure 4.5-1 shows some mode patterns of the Laguerre–Gaussian mode. It is seen from this figure that the mode number m and $2l$ correspond to the number of nodes in radial and azimuthal directions in the mode pattern. Figure 4.5-2 shows the dispersion curve of a DI round fiber. In this figure, the solid line indicates the exact value calculated by the stratified multilayer matrix method,[7] and the dashed line shows the Hermite–Gaussian and Laguerre–Gaussian modes. Since the Hermite–Gaussian and Laguerre–Gaussian modes are derived under the assumption that the field is well confined in the core region, the lines do not coincide near the cutoff. However, when the mode is far from the cutoff, the Laguerre–Gaussian mode gives a good approximation.

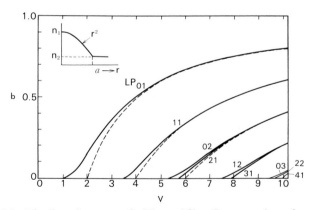

FIGURE 4.5-2. The dispersion curves of a DI round fiber. Curves are shown for exact analysis (solid lines) and for the Hermite–Gaussian or Laguerre–Gaussian mode (dashed lines). (After Tanaka and Suematsu.[7])

4.6 SINGLE-MODE WAVEGUIDE

In the previous sections of this chapter, we dealt with the multimode waveguides and fibers. However, when the width of the waveguide or core radius of the fiber is relatively small and of the same order of magnitude as the wavelength, this waveguide can maintain only the fundamental mode. Such

waveguides, called single-mode waveguides, are generally used as the active
layer of semiconductor lasers and as the waveguide of optical devices.

A. SINGLE-MODE CONDITION OF PLANAR WAVEGUIDES

The simplest structure of a single-mode waveguide is the step-index single-
mode planar waveguide, shown in Fig. 4.6-1(a). As discussed in Section 4.2, we
can use Eq. (4.2-24) to obtain the single-mode condition of a step-index planar
waveguide:

$$V < \pi/2. \tag{4.6-1}$$

However, the single-mode condition of the distributed index planar waveguide shown in Fig. 4.6-1(b) is rather complicated and requires an advanced
mathematical technique. An outline of this technique, the cutoff theory[8] for
distributed index planar waveguides, will be discussed.

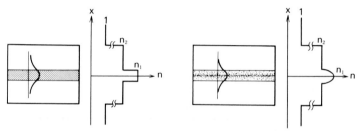

FIGURE 4.6-1.   Single-mode planar waveguides. Shown are (a) a step-index single-mode planar
waveguide and (b) a distributed-index single-mode planar waveguide.

Let us consider an arbitrary index distribution expressed by

$$n^2(x) = \begin{cases} n_1^2[1 - 2\Delta f(x)], & |x| \le a, \\ n_2^2, & |x| > a, \end{cases} \tag{4.6-2}$$

where $2\Delta = (n_1^2 - n_2^2)/n_1^2$ and $f(x)$ is assumed to be symmetric to $x = 0$. The
wave equation and boundary conditions of odd TE modes at cutoff are given
by

$$d^2 E_y/d\tilde{x}^2 + V_c^2[1 - f(\tilde{x})]E_y = 0, \tag{4.6-3}$$

$$E_y(0) = 0, \tag{4.6-4}$$

$$dE_y/d\tilde{x}\Big|_{x=1} = 0, \tag{4.6-5}$$

where

$$\tilde{x} = x/a \tag{4.6-6}$$

and V_c is the cutoff V value. Since Eqs. (4.6-3)–(4.6-5) constitute a Sturm–Liouville-type boundary problem, these conditions can be transformed into a symmetric kernel integral equation as

$$y(\tilde{x}) = V_c^2 \int_0^1 K(\tilde{x}, \tilde{x}') y(\tilde{x}') d\tilde{x}', \qquad (4.6\text{-}7)$$

where

$$y(\tilde{x}) = \sqrt{1 - f(\tilde{x})} E_y(\tilde{x}), \qquad (4.6\text{-}8)$$

$$K(\tilde{x}, \tilde{x}') = \begin{cases} \sqrt{1 - f(\tilde{x})}\,\tilde{x}'\sqrt{1 - f(\tilde{x}')}, & 0 \le \tilde{x} < \tilde{x}' \le 1, \\ \sqrt{1 - f(\tilde{x})}\,\tilde{x}\sqrt{1 - f(\tilde{x}')}, & 0 \le \tilde{x}' < \tilde{x} \le 1. \end{cases} \qquad (4.6\text{-}9)$$

The cutoff V value V_c of the TE_1 mode is obtained as the minimum eigenvalue of Eq. (4.6-7). The simplest cutoff formula for the TE_1 mode is given by[8]

$$V_c \simeq \left[2 \int_0^1 (1 - f(\tilde{x})) \left\{ \int_0^{\tilde{x}} \tilde{x}'(1 - f(\tilde{x}')) d\tilde{x}' \right\} d\tilde{x} \right]^{-1/4} \qquad (4.6\text{-}10)$$

When the index distribution is expressed by an α-power law, substitution of $f(\tilde{x}) = \tilde{x}^\alpha$ into Eq. (4.6-10) gives

$$V_c = \left[\frac{1}{6} + \frac{1}{\alpha + 2} - \frac{3}{\alpha + 3} + \frac{4}{3} \frac{1}{\alpha + 4} \right]^{-1/4} \qquad (4.6\text{-}11)$$

Therefore, a distributed-index planar waveguide whose index distribution is expressed by an α–power law maintains only a TE_0 mode when V is smaller than the V_c given by Eq. (4.6-11). Figure 4.6-2 shows the cutoff V value V_c of the TE_1 mode versus the exponent α of the α–power law index distribution.

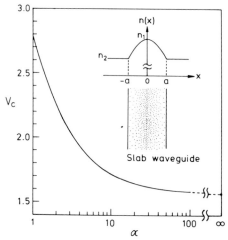

FIGURE 4.6-2.   Single-mode condition of a distributed-index planar waveguide.

B. SINGLE-MODE CONDITION OF THREE-DIMENSIONAL WAVEGUIDE

Single-mode waveguides used for optical devices usually have three-dimensional configurations, as shown in Fig.4.6-3. To obtain the exact single-mode condition of these three-dimensional waveguides, several numerical analysis methods are needed.[9-13] However, these methods require computer calculation and serve no analytical prospects. In this subsection, an approximated analytical method called the equivalent-index method will be introduced.[14]

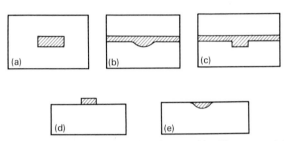

FIGURE 4.6-3. Several types of three-dimensional waveguides. Shown are a (a) rectangular; (b) lens-like; (c) rib; (d) strip; and (e) diffused waveguide.

First we shall consider a planar waveguide, whose core thickness varies in the y direction as shown in Fig.4.6-4(a). The field in the waveguide is assumed to be expressed by

$$E_y(x, y) = \psi(y) \cos(\kappa(y)x) \exp[j(\omega t - \beta z)]. \qquad (4.6\text{-}12)$$

Here $\kappa(y)$ can be considered a local propagation constant in x direction of a step-index planar waveguide whose core thickness is $d(y)$. Therefore $\kappa(y)$ is

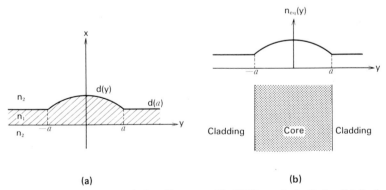

FIGURE 4.6-4.   (a) Cross section of a lens-like waveguide. (b) The equivalent index distribution.

determined by solving a characteristic equation

$$\tan[\kappa(y)\,d(y)/2] = \sqrt{k_0^2(n_1^2 - n_2^2) - \kappa^2(y)}/\kappa(y)$$
$$= \sqrt{V^2(y) - \kappa^2(y)\,d^2(y)/4}/[\kappa(y)\,d(y)/2] \quad (4.6\text{-}13)$$

for an even TE mode.

If we assume that $\kappa(y)$ varies slowly compared with $\psi(y)$, the term

$$\left\{ \frac{d^2\kappa(y)}{dy^2}\,\psi(y)\sin[\kappa(y)x] + \left[\frac{d\kappa(y)}{dy}\right]^2 \psi(y)\cos[\kappa(y)x] \right.$$
$$\left. + 2\,\frac{d\kappa(y)}{dy}\frac{d\psi(y)}{dy}\sin[\kappa(y)x] \right\}$$

can be neglected, and we obtain the following equation by substituting Eq. (4.6-12) into Eq. (4.2-6):

$$\frac{d^2\psi}{dy^2} + (k_0^2 n_1^2 - \kappa^2(y) - \beta^2)\psi(y) = 0, \quad (4.6\text{-}14)$$

where the gradient-index term was neglected. Equation (4.6-14) shows that the thickness variation $d(y)$ can be considered an equivalent index distribution $n_{eq}(y)$ expressed by

$$n_{eq}^2(y) = n_1^2 - [\kappa^2(y)/k_0^2]. \quad (4.6\text{-}15)$$

The variation of $d(y)$ brings about the variation of $\kappa(y)$ in Eqs. (4.6-13)–(4.6-15). However, since the variation of $d(y)$ also causes the variation of V value in the x direction as a function of y, we can obtain $\kappa(y)\,d(y)$ from Fig. 4.2-3 instead of solving Eq. (4.6-13). Therefore, the single-mode condition of many three-dimensional waveguides can be evaluated by replacing the thickness variation with the equivalent index distribution and by utilizing Eq. (4.6-10) or Eq. (4.6-11).

For example, we can evaluate the waveguide width $2a$ required for single transverse mode propagation by

$$2a_s = \frac{2V_c}{k_0\sqrt{n_{eq}^2(0) - n_{eq}(a)}} \quad (4.6\text{-}16)$$

where V_c is determined from $n_{eq}(y)$ through Eq. (4.6-10) or Eq. (4.6-11). Here the equivalent index n_{eq} is related to $b(V)$ which is defined by Eq. (4.2-22) and is given by

$$n_{eq}^2(y) = (n_1^2 - n_2^2)b(V) + n_2^2. \quad (4.6\text{-}17)$$

When the $V-b$ dispersion curve for the fundamental mode shown in Fig. 4.2-3 can be approximated by a linear relation in the range of $0 \leq V \leq \pi/2$, that is,

$$b(V) \simeq 0.4V(y) = 0.2k_0n_1 \sqrt{2\Delta} \, d(y) \tag{4.6-18}$$

the single-mode waveguide width $2a$ is simply given by

$$2a_s = \frac{2V_c}{\{0.2k_0^3n_1^3(2\Delta)^{3/2}[d(0) - d(a)]\}^{1/2}}, \tag{4.6-19}$$

since $n_{eq}^2(y) = 0.2k_0n_1^3(2\Delta)^{3/2} d(y) + n_2^2$. Note that the shape of the waveguide is included only into the determination of cut-off frequency V_c.

C. SINGLE-MODE CONDITION OF ROUND FIBER

Since the second-highest-order modes of round fibers are TE_{01}, TM_{01}, and HE_{21} modes, the cutoff V value of these modes determines the single-mode condition of round fibers. In the case of step-index round fiber, putting $\beta \rightarrow k_0n_2$ ($w \rightarrow 0$) in Eqs. (4.4-30) and (4.4-31) with $l = 0$, we obtain

$$J_0(u) = 0. \tag{4.6-20}$$

Therefore, the cutoff V value V_c of the TE_{01} and TM_{01} modes is given by the first zero of the 0th-order Bessel function $V_c = 2.4048256$. The cutoff V_c of the HE_{21} mode is slightly larger than that of the TE_{01} and TM_{01} modes.[15,16]

However, the cutoff problem of distributed-index round fiber is rather complicated and requires an advanced mathematical technique used in the derivation of Eq. (4.6-10). The detailed derivation process[8] is omitted here and only the result is given. When the index distribution is expressed by

$$n^2(\rho) = \begin{cases} n_1^2[1 - 2\Delta f(\rho)], & \rho \leq 1, \\ n_2^2, & \rho > 1, \end{cases} \tag{4.6-21}$$

where $2\Delta = (n_1^2 - n_2^2)/n_1^2$, $\rho = r/a$, and $f(\rho)$ is an arbitrary function, the simplest cutoff formula of TE_{01} mode is given by

$$V_c \simeq 1/\sqrt[4]{T_2}, \tag{4.6-22}$$

where

$$T_2 = \frac{1}{2} \int_0^1 \int_0^t \frac{s^3}{t} [1 - f(s)][1 - f(t)] \, ds \, dt. \tag{4.6-23}$$

When $f(\rho) = \rho^\alpha$, Eqs. (4.6-18) and (4.6-19) become

$$V_C \simeq \left[\frac{1}{32} + \frac{1}{8} \frac{1}{(\alpha + 2)} - \frac{1}{4} \frac{1}{(\alpha + 4)} - \frac{1}{2} \frac{1}{(\alpha + 4)^2} \right]^{-1/4} \tag{4.6-24}$$

Although Eqs. (4.6-22) and (4.6-23) have an error of 1 to 2%, these equations are useful for rough evaluation because they explicitly include the index distribution. More accurate formulas are given in References 8 and 16. Figure 4.6-5 shows the single-mode condition of α-power law fibers whose index

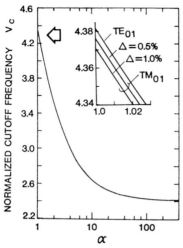

FIGURE 4.6-5. Single-mode condition of distributed-index round fiber. (See text for description. After Kokubun and Iga.[8,16])

distribution function is expressed by $f(\rho) = \rho^{\alpha}$. A similar result was obtained by Gambling *et al.* by means of the Taylor series expansion method.[17] The inset in Fig. 4.6-5 shows the difference of V_c's of TE_{01} and TM_{01} modes calculated by a more accurate analysis.[16]

Finally, we consider the effect of a center dip on V_c. The index distribution including a center dip is expressed by a Lorentian function

$$f(\rho) = (W/a)^2(1 - \rho^2)/[(W/a)^2 + \rho^2], \qquad (4.6\text{-}25)$$

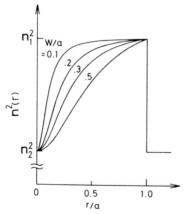

FIGURE 4.6-6. The Index distribution including a Lorentzian-form center dip of width W.

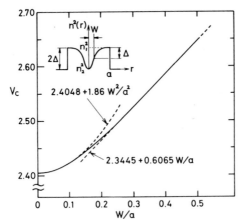

FIGURE 4.6-7.   The single-mode condition of a fiber with Lorentzian-form center dip of width
W. (After Kokubun and Iga.[15])

where a is the core radius and W corresponds to the width of center dip, as
shown in Fig. 4.6-6. The single-mode condition of this type of fiber is shown in
Fig. 4.6-7.[16]

REFERENCES

1.   H. G. Unger, *Arch. Elek, Ubertragung* **19**, 189 (1965).
2.   Y. Itaya, Y. Suematsu, S. Katayama, K. Kishino, and S. Arai, *Jpn. J. Appl. Phys.* **18**, 1795 (1979).
3.   L. I. Schiff, *Quantum Mechanics*, 3rd ed. McGraw–Hill, New York, 1968.
4.   J. A. Stratton, *Electromagnetic Theory*, McGraw–Hill, New York, 1941.
5.   Snitzer, *J. Opt. Soc. Am.* **51**, 491 (1961).
6.   D. Gloge, *Appl. Opt.* **10**, 2252 (1971).
7.   T. Tanaka and Y. Suematsu, *Trans. IECE Jpn,* **E59**, 1 (1976).
8.   Y. Kokubun and K. Iga, *J. Opt. Soc. Am.* **70**, 36 (1980).
9.   C. Yeh, K. Ha, S. B. Dong, and W. P. Brown, *Appl. Opt.* **18**, 149 (1979).
10. J. E. Goell, *Bell Syst. Tech. J.* **48**, 2133 (1969).
11. K. Yasuura, K. Shimohara, and T. Miyamoto, *J. Opt. Soc. Am.* **70**, 183 (1980).
12. M. Geshiro, Y. Masaki, M. Ohtaka, M. Matsuhara, and N. Kumagai, *Trans. IECE Jpn.* **60-C**, 256 (1977).
13. E. A. J. Marcatili, *Bell Syst. Tech. J.* **48**, 2071 (1969).
14. K. Iga, *Appl. Opt.* **19**, 2940 (1980).
15. Y. Kokubun and K. Iga, *J. Opt. Soc. Am.* **70**, 388 (1980).
16. Y. Kokubun and K. Iga, *Radio Sci.* **17**, 43 (1982).
17. W. A. Gambling, D. N. Payne, and H. Matsumura, *Electron Lett.* **13(5)**, 139 (1977).

CHAPTER 5

Image Transmission and Transformation

5.1 OPENING REMARKS

As mentioned in Chap. 3, in the earlier history of lightwave transmission a DI medium was considered a promising device among continuously focusing light guides having geometries of fibers (see, for example, Kapany[1]) or slabs (see, for example, Suematsu[2]).

Some types of lenslike[3] and square-law media[4] have been proposed and studied for light-beam waveguides for laser communication systems. Various types of gas lenses[5,6] and focusing glass fibers[7] provide examples of such media, whose dielectric constant is gradually graded with a square law with regard to the distance from the center axis. As for optical characteristics of such media, it is known that a Hermite–Gaussian light beam is guided along the axis of the lenslike medium and moreover, images are transformed with a definite transform law that not only maintains the information of their intensity distribution but also of their phase relation. This is thought to be one of the significant characteristics of gas lenses and focusing glass fibers, a characteristic different from that of a step-index glass fiber.[8]

Various authors have reported on imaging properties of DI media. Aoki and Suzuki[9] investigated the imaging property of a gas lens, where the lens formula and optical transfer function were obtained on the basis of geometrical optics, and some imaging experiments were made using a flow-type gas lens. In a paper describing the optical characteristics of a light-focusing fiber guide (SELFOC), Uchida et al.[10] mentioned the experimental imaging property and measured the resolving power of the SELFOC lens. In each of these papers, imaging properties of DI media are interpreted in terms of geometrical optics.

When a DI medium is applied to coherent optics such as interferometry and holography, however, it is important that a two-dimensional system theory based on wave optics[11,12] be introduced into the treatment of transforms by an optical system with a DI medium. In this chapter we introduce such a theory which applies an integral transform associated with a DI medium into

75

the system theory of optics (partly discussed in Reference 13). This will enable us to learn not only about an imaging condition but also about some types of transform representations. Also in this chapter, the results of experiments done on the imaging properties of the optical system using SELFOC fibers are compared with the theory. In addition, discussions on the pass band of spatial frequencies are presented.

As a good example of different transmission of images, we introduce step-index imaging in the last part of this chapter. This corresponds to "pulse-modulated" imaging, whereas the DI imaging is said to be "analog modulated."

5.2 CONDITION FOR DISTRIBUTED-INDEX IMAGING

In this chapter, the index-profile of a DI medium is expressed by the following equation, first proposed by Suematsu and Iga:[14]

$$n^2(r)\begin{cases} = n^2(0)[1 - (gr)^2 + h_4(gr)^4 + h_6(gr)^6 + \cdots] & r < a \\ = n_2^2 & r > a \end{cases} \quad (5.2\text{-}1)$$

where $n(0)$ is the index at the center axis, a the core radius, and n_2 the index out of the core as shown in Fig. 5.2-1.

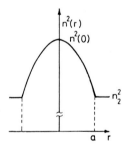

FIGURE 5.2-1.   Expression of index distribution. (After Iga, Hata, Kato, and Fukuyo.[19])

If we express the transverse field component by $\exp(j\omega t - j\beta z)$, the function ψ for the index profile given by Eq. (1) is given approximately by the scalar wave equation

$$\frac{1}{r}\frac{d}{dr}\left(r\frac{d\psi}{dr}\right) + \frac{d^2\psi}{d\theta^2} + k_0^2 n^2(r)\psi = \beta^2\psi \quad (5.2\text{-}2)$$

in (r, θ, z) cylindrical coordinates, where $k_0 = 2\pi/\lambda$.

The normal modes associated with a square-law medium, using the first two terms of Eq. (5.2-1), are known to be Hermite–Gaussian[12] or Laguerre–Gaussian[13] functions.

The characteristic spot size w_0 of the fundamental mode is given by

$$w_0 = a/(V)^{1/2}, \tag{5.2-3}$$

where the normalized frequency V is written as

$$V = k_0 n(0) a (2\Delta)^{1/2}, \tag{5.2-4}$$

with $\Delta = [n(0) - n(a)]/n(0)$. In the usual GRIN lenses,[1,2,4,5,15] V is larger than 3000, since we have $\lambda = 0.5 \, \mu m$, $n(0) = 1.5$, $a = 0.5 \, mm$, and $\Delta = 5\%$. The characteristic spot size w_0 is therefore smaller than the core radius a by a factor of 50 to 100. The ratio $w_0/a = (V)^{1/2}$ is a measure to indicate whether or not we can use this gradient medium as an imaging lens, because a sinusoidal ray trace is distorted if w_0/a is not small. The propagation constant β_{nm} associated with the index profile given by Eq. (5.2-1), including higher-order terms, is obtained by a perturbation method[14] and expressed in terms of a series expansion in powers of $g/k(0)$, where $k(0) = k_0 n(0)$, as follows:

$$\frac{\beta_{lm}}{k(0)} = 1 - (2l + m + 1)\left[\frac{g}{k(0)}\right] + \frac{1}{2}\left\{h_4\left[\frac{3}{2}(2l + m + 1)^2\right.\right.$$

$$\left.\left. + \frac{1}{2}(1 - m^2)\right] - (2l + m + 1)^2\right\} \times \left[\frac{g}{k(0)}\right]^2 + O\left\{\left[\frac{g}{k(0)}\right]^3\right\}. \tag{5.2-5}$$

We should note that $g/k(0)$ is of the order of 10^{-4} to 10^{-3}. If $m = 0$, Eq. (5.2-5) is associated with the radially symmetric mode that corresponds to meridional rays.

If we calculate the group velocity v_g by differentiating $\beta_{n(0)}$ with respect to ω,[15] we see that the minimum dispersion condition is $h_4 = \frac{2}{3}$ and the same result obtained from the WKB method.[19]

5.3 MODE EXPANSION METHOD

In Chapter 4, we obtained the eigen modes for a parabolically distributed-index waveguide. The nature of the eigen mode is such that it propagates along the waveguide without changing its transverse field distribution and with the associated propagation constant when the particular mode is excited at the incident position. If any arbitrary distribution of the field is excited at the input end, the power of that field is distributed into eigen modes, and partially into radiation modes, and each mode carries a distributed power of a light wave with the group velocity of that mode.

When the field $f(x, y, 0)$ is incident at $z = 0$ of the DI waveguide, as shown in Fig. 5.3-1, $f(x, y, 0)$ is expanded by the eigen modes since they constitute the complete set:

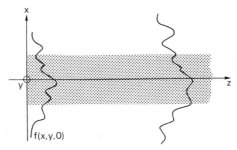

FIGURE 5.3-1.   Transformation of spatial information.

$$f(x, y, 0) = \sum_{p=0,q=0}^{\infty} a_{pq} u_p(x, w_{01}) u_q(y, w_{02}).\qquad(5.3\text{-}1)$$

If we integrate both sides after multiplying by $u_{p'}(x, w_{01})u_{q'}(y, w_{02})$, we have

$$a_{p'q'} = \int_{-\infty}^{\infty} \int_{-\infty}^{\infty} f(x, y, 0)\, u_{p'}(x, w_{01})\, u_{q'}(y, w_{02})\, dx\, dy.\qquad(5.3\text{-}2)$$

Here we have used the orthonormality of the eigen function:

$$\int_{-\infty}^{\infty} u_p(x, w_{01})\, u_{p'}(x, w_{01})\, dx = \delta_{pp'}.\qquad(5.3\text{-}3)$$

Since each mode propagates with the propagation constant β_{pq} given by Eq. (5.2-5), the field is expressed by the equation

$$f(x, y, z) = \sum_{p=0,q=0}^{\infty} a_{pq} u_p(x, w_{01}) u_q(y, w_{02}) \exp[-j\beta_{pq} z],\qquad(5.3\text{-}4)$$

after propagating by distance z. We have assumed that no mode conversion or mode mixing occurs during the propagation. It is generally difficult to find the distribution of the field by calculating the sum of Eq. (5.3-4) because of the phase difference among modes.

But fortunately, perhaps miraculously, the summation of Eq. (5.3-4) can be performed analytically if we postulate the idealized condition on the waveguide, which is actually not so different from the real one.[17] We try to show this in the following. By substituting Eqs. (5.2-5) and (5.3-2) into Eq. (5.3-4), we obtain

$$f(x, y, z) = \exp(-jkz)\sum_{p,q} \int_{-\infty}^{\infty}\int_{-\infty}^{\infty} f(x', y', 0)u_p(x', w_{01})u_q(y', w_{02})\, dx'\, dy'$$
$$\times\, u_p(x, w_{01})u_q(y, w_{02}) \exp[jg_1(p + \tfrac{1}{2})z + jg_2(q + \tfrac{1}{2})z]\qquad(5.3\text{-}5)$$

After interchanging the integral and summation, Eq. (5.3-5) reduces to

$$f(x, y, z) = \frac{j}{\lambda z} \sqrt{\frac{g_1 z}{\sin g_1 z} \frac{g_2 z}{\sin g_2 z}} \exp(-jkz)$$

$$\times \iint dx' \, dy' \, f(x', y', 0) \, K(x, x'; y, y'), \qquad (5.3\text{-}6)$$

where

$$K(x, x'; y, y') = \exp\left[-\frac{j}{2w_{01}^2} \cot(g_1 z)(x^2 - 2xx' \sec g_1 z + x'^2) \right]$$

$$\times \exp\left[-\frac{j}{2w_{02}^2} \cot(g_2 z)(y^2 - 2yy' \sec g_2 z + y'^2) \right]. \qquad (5.3\text{-}7)$$

Here we use Mehler's formula;

$$\sum_{n=0}^{\infty} \frac{(\frac{1}{2}\zeta)^n}{n!} H_n(x)H_n(x') = (1 - \zeta^2)^{-1/2} \exp\left[\frac{2xx'\zeta - (x^2 + x'^2)\zeta^2}{1 - \zeta^2} \right]. \qquad (5.3\text{-}8)$$

When $g_1 = g_2 = g$, the circularly cylindrical coordinate is conveniently expressed. It is, of course, possible to reach a result similar to Eq. (5.3-6) if we expand the field in terms of Laquerre–Gaussian modes. However, we can (at least) obtain the closed form by transforming the coordinates. If we put

$$x = r \cos \theta, \qquad y = r \sin \theta \qquad (5.3\text{-}9)$$

and transform Eq. (5.3-6), we obtain

$$f(r, \theta, z) = \frac{j}{\lambda z} \frac{gz}{\sin gz} \exp(-jkz)$$

$$\times \iint r' \, dr' \, d\theta' \, f(r', \theta', 0) \, K(r, \theta; r', \theta'), \qquad (5.3\text{-}10)$$

where the kernel is given by

$$K(r, \theta; r', \theta')$$

$$= \exp\left\{ -\frac{j}{2w_0^2} \cot gz \, [r^2 - 2rr' \cos(\theta - \theta') \sec gz + r'^2] \right\}. \qquad (5.3\text{-}11)$$

Here, $w_{01} = w_{02} = w_0$ with

$$w_0^2 = 1/gk. \qquad (5.3\text{-}12)$$

Now let us investigate the result of this calculation. Generally speaking, if we have the integral expression

$$f(x, y, z) = \iint_S f(x', y', 0) \, G(x, y; x', y') \, dx' \, dy', \qquad (5.3\text{-}13)$$

the function $G(x, y; x', y')$ is a Green's function, associated with the 2-D impulse response of the optical system. From another viewpoint, it may be called a modified Fresnel–Kirchhoff integral associated with the DI rod lens.

An interesting case occurs when the focusing effect is very weak. When $g \to 0$, Eqs. (5.3-6) and (5.3-10) approach

$$f(x, y, z) = \frac{j}{\lambda z} \exp(-jkz) \int \int dx' \, dy' \, f(x', y', 0)$$

$$\times \exp\left\{ -\frac{jk}{2z} [(x - x')^2 + (y - y')^2] \right\} \qquad (5.3\text{-}14)$$

or

$$f(r, \theta, z) = \frac{j}{\lambda z} \exp(-jkz) \int \int r' \, dr' \, d\theta \, f(r', \theta', 0)$$

$$\times \exp\left\{ -\frac{jk}{2z} [r'^2 - 2rr' \cos(\theta - \theta') + r^2] \right\} \qquad (5.3\text{-}15)$$

These mathematical expressions of diffraction are well-known Fresnel–Kirchhoff integrals. In DI imaging, as we have seen, the image is transmitted by being distributed to guided modes. There, the phase relationship among modes is not lost, and is recombined with the definite superimposing law. Therefore, coherent imaging can be performed.

5.4 DISTRIBUTED-INDEX IMAGING

A. INTEGRAL TRANSFORMS

The transverse distribution of the dielectric constant or squared refractive index (associated here with a DI medium of interest) is expressed by the equation

$$n^2(x, y) = n^2(0)[1 - (g_1 x)^2 - (g_2 y)^2 + h(x, y)], \qquad (5.4\text{-}1)$$

where $n(0)$ is the dielectric constant on the center axis and $h(x, y)$ contains higher terms on x and y far smaller than $(g_1 x)^2$ and $(g_2 y)^2$. In this case, the propagation constant β_{pq} associated with an Hermite–Gaussian beam[3] may be assumed to be given simply by the equation

$$\beta_{pq} = k(0) - (p + \tfrac{1}{2})g_1 - (q + \tfrac{1}{2})g_2, \qquad (5.4\text{-}2)$$

where $k(0) = \omega\sqrt{\varepsilon(0)\mu_0}$ with the angular optical frequency $\varepsilon(0) = \varepsilon_0 n(0)$, and μ_0 denotes the permeability of the medium and p (or q) the pth (or qth) mode in the x (or y) direction. This assumption is not so far-fetched, because higher-order terms on p and q in β_{pq} can be compensated for in the ideal case by higher-order terms[14,15] from the index distribution $h(x, y)$, which has higher terms on x and y. It is necessary, of course, that the distortion due to these higher terms be taken into consideration. This will be discussed in another section.

In the ideal case mentioned above, the transverse electric field $\phi_1(x_1, y_1)$ in front of the DI medium of length b in Fig. 5.4-1 is transformed by the integral transform expressed by the equation[17]

$$\phi_2(x_2, y_2) = \frac{j}{\lambda b} \exp[-jk(0)b] \sqrt{\frac{g_1 b}{\sin g_1 b} \frac{g_2 b}{\sin g_2 b}}$$

$$\times \exp\left[j\frac{1}{2}\left(\frac{x_2}{w_{01}}\right)^2 \tan g_1 b + j\frac{1}{2}\left(\frac{y_2}{w_{02}}\right)^2 \tan g_2 b\right]$$

$$\times \int\int dx_1\, dy_1 \phi_1(x_1, y_1) \exp[-j\beta_1(x_1 - x_2 \sec g_1 b)^2$$

$$- j\beta_2(y_1 - y_2 \sec g_2 b)^2], \tag{5.4-3}$$

where w_{0i} and $\beta_i(i = 1, 2)$ are defined by the equations

$$w_{0i}^2 = \frac{1}{k(0)g}, \tag{5.4-4}$$

$$\beta_i = \frac{k(0)}{2b} \frac{g_i b}{\tan g_i b}, \qquad i = 1, 2. \tag{5.4-5}$$

Next we assume that the medium has losses and that, for simplicity, the pth and qth Hermite–Gaussian modes in the x and y direction attenuate by the rate $\Gamma_1(p + \frac{1}{2})$ and $\Gamma_2(q + \frac{1}{2})$, respectively. When we expand $\phi_1(x_1, y_1)$ in terms

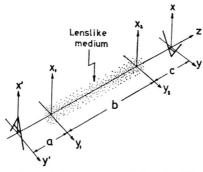

FIGURE 5.4-1. An image transmission system consisting of a DI medium. (After Iga, Hata, Kato, and Fukuyo.[19])

of the Hermite–Gaussian modes, the expansion coefficient a_{pq} associated with the pq mode is assumed to be

$$a_{pq}(b) = a_{pq}(0) \exp[-(p + \tfrac{1}{2})\Gamma_1 b - (q + \tfrac{1}{2})\Gamma_2 b. \qquad (5.4\text{-}6)$$

In this case, the integral in Eq. (5.4-3) is modified as shown by the equation

$$
\begin{aligned}
\phi_2(x_2, y_2) = {} & \frac{j}{\lambda b} \exp[-jk(0)b] \cdot \frac{1}{\sqrt{\sin(g_1 + j\Gamma_1)b \, \sin(g_2 + j\Gamma_2)b}} \\
& \times \exp\left[j\frac{1}{2}\left(\frac{x_2}{w_{02}}\right)^2 \tan(g_1 + j\Gamma_1)b + j\frac{1}{2}\left(\frac{w_{02}}{y_2}\right)^2 \tan(g_2 + j\Gamma_2)b \right] \\
& \times \iint dx_1 \, dy_1 \, \phi_1(x_1, y_1) \exp[-j\tilde{\beta}_1(x_1 - x_2 \sec(g_1 + j\Gamma_1)b)^2 \\
& \quad -j\tilde{\beta}_2(y_1 - y_2 \sec(g_2 + j\Gamma_2)b)^2],
\end{aligned} \qquad (5.4\text{-}7)
$$

where the complex parameter $\tilde{\beta}_i$ is expressed by the equation

$$\tilde{\beta}_i = \frac{k(0)}{2b} \frac{g_i b}{\tan(g_i + j\Gamma_i)b}, \qquad i = 1, 2. \qquad (5.4\text{-}8)$$

Physically this situation is associated with the square-law loss mechanism.[18]

B. System Theory

In this section we shall proceed to the transforms in a two-dimensional system consisting of free space–lenslike medium–free space, as shown in Fig. 5.4-1. The transverse electric field component $\phi(x', y')$ in the (x', y') plane is known to be transformed by the Fresnel diffraction law and given by the two-dimensional convolution[12]

$$\phi_1(x_1, y_1) = \frac{j}{\lambda_0 a} \exp[-jk_0 a][f(x_1, y_1) ** \exp[-j\alpha(x_1^2 + y_1^2)], \qquad (5.4\text{-}9)$$

$$\alpha = k_0/2a, \qquad (5.4\text{-}10)$$

where $**$ denotes the two-dimensional convolution defined as

$$[P(x, y) ** Q(x, y)] = \iint dx' \, dy' \, P(x, y) \, Q(x - x', y - y'). \qquad (5.4\text{-}11)$$

If we know $\phi_2(x_2, y_2)$ in the (x_2, y_2) plane, the final transformed electric field $g(x, y)$ in the (x, y) plane is also given by

$$g(x, y) = j/\lambda_0 c \, \exp[-jk_0 c][\phi_2(x, y) ** \exp[-j\gamma(x^2 + y^2)], \qquad (5.4\text{-}12)$$

$$\gamma = k_0/2c. \qquad (5.4\text{-}13)$$

The final distribution $g(x, y)$ is obtained by substituting Eqs. (5.4-3) and (5.4-9) into Eq. (5.4-12) and expressed by the equation[13,19]

$$g(x, y) = C \exp[-j\beta_1\gamma/(\beta_1\gamma)x^2 - j\beta_2\gamma/(\beta_2 + \gamma)y^2]$$

$$\times \int\int dx_1\, dy_1\, F_1(2\alpha x_1, 2\alpha y_1) \exp[jXx_1 + jYy_1]$$

$$\times \exp\{-j[\alpha + \beta_1 - \beta_1^2 \sec^2(g_1 b)/(\beta_1 + \gamma)]x_1^2\}$$

$$\times \exp\{-j[\alpha + \beta_2 - \beta_2^2 \sec^2(g_2 b)/(\beta_2 + \gamma)]y_1^2\}, \qquad (5.4\text{-}14)$$

where

$$C = (j/\lambda_0 a)(1/2\lambda b)(j/\lambda_0 c) \exp[-j(k_0 a + k(0)b + k_0 c)]$$

$$\times [(g_1 b)(g_2 b)/\sin(g_1 b)\sin(g_2 b)]^{1/2}[(\beta_1 + \gamma)(\beta_2 + \gamma)]^{-1/2}, \quad (5.4\text{-}15)$$

$$X = \frac{2\beta_1\gamma \sec g_1 b}{\beta_1 + \gamma}\, x, \qquad (5.4\text{-}16)$$

$$Y = \frac{2\beta_2\gamma \sec g_2 b}{\beta_2 + \gamma}\, y, \qquad (5.4\text{-}17)$$

and $F_1(u, v)$ is defined as the Fourier transform of $f_1(x, y)$ as

$$f_1(x, y) = f(x, y) \exp[-j\alpha(x^2 + y^2)], \qquad (5.4\text{-}18)$$

$$F_1(u, v) = \int\int dx\, dy\, f_1(x, y) \exp(jux + jvy). \qquad (5.4\text{-}19)$$

The inverse transform of Eq. (5.4-19) is given by

$$f_1(x, y) = (1/2\pi)^2 \int\int du\, dv\, F_1(u, v) \exp(-jxu - jyv). \qquad (5.4\text{-}20)$$

C. IMAGING CONDITION

We shall now examine Eq. (5.4-14), which describes the field transformed by the two-dimensional system, as shown in Fig. 5.4-1. This equation reduces to a simple Fourier transform of $F_1(2\alpha x_1, 2\alpha y_1)$ when the last two exponents vanish, that is,

$$\alpha + \beta_i - \beta_i^2 \sec^2(\beta_i b)/(\beta_i + \gamma) = 0, \qquad i = 1, 2, \qquad (5.4\text{-}21)$$

and the limits of the integration are assumed to extend to infinity. By substituting Eqs. (5.4-5), (5.4-10), and (5.4-13) into Eq. (5.4-21), we obtain

$$\left(\frac{k_0}{a} + \frac{k(0)}{b}\frac{g_i b}{\tan g_i b}\right)\left(\frac{k_0}{c} + \frac{k(0)}{b}\frac{g_i b}{\tan g_i b}\right) = \left(\frac{k(0)}{b}\right)^2 \left(\frac{g_i b}{\sin g_i b}\right)^2 \qquad i = 1,2.$$

$$(5.4\text{-}22)$$

In this case, the integral in Eq. (5.4-14) can be of the form of Eq. (5.4-20), which reduces to be readily carried out by using the relationship

$$g(x, y) = C(\pi/\alpha)^2 \exp\left[-j\frac{\gamma}{\alpha}\frac{\alpha\beta_1 + \beta_1\gamma + \gamma\alpha}{\beta_1 + \gamma}x^2 - j\frac{\gamma}{\alpha}\frac{\alpha\beta_2 + \beta_2\gamma + \gamma\alpha}{\beta_2 + \gamma}y^2\right]$$

$$\times f\left(\frac{x}{m_1}, \frac{y}{m_2}\right), \qquad (5.4\text{-}23)$$

$$m_i = -\alpha(\beta_i + \gamma)/(\beta_i\gamma \sec g_i b). \qquad (5.4\text{-}24)$$

Equation (5.4-23) shows that $g(x, y)$ is the image of $f(x, y)$, whose dimension is multiplied by m_1 or m_2 in the x or y direction. For this reason, Eq. (5.4-22) is read as the imaging condition and m_i the multiplication factor. If we write $k(0)/k_0 = n(0)$, then the imaging condition Eq. (5.4-22) is reduced to

$$\left(\frac{1}{a} + \frac{n(0)g}{\tan gb}\right)\left(\frac{1}{c} + \frac{n(0)g}{\tan gb}\right) = \frac{n^2(0)g^2}{\sin gb}. \qquad (5.4\text{-}25)$$

After some manupilations we obtain

$$\tan(gb) = \frac{(a + c)n(0)g}{n^2(0)g^2ac - 1}. \qquad (5.4\text{-}26)$$

In the same way, magnification m is given by

$$m = -\left(\frac{1}{n(0)ga}\sin gb + \frac{c}{a}\cos gb\right) \qquad (5.4\text{-}27)$$

The distance c is related to a and given by

$$c = \frac{1}{n(0)g}\frac{\tan gb + n(0)ga}{n(0)ga \tan gb - 1}. \qquad (5.4\text{-}28)$$

Then m is finally written as

$$m = \frac{1}{\cos gb - n(0)ga \sin gb}. \qquad (5.4\text{-}29)$$

The same results will be given by Eqs. (6.4-13) and (6.4-14), respectively, with the help of the matrix method.

D. EXPERIMENTAL VERIFICATION

To verify the results of the previous analysis, some experiments on the transmission of images were done by making use of the SELFOC fiber.[7] The

TABLE 5.4-I

SAMPLES OF SELFOC FIBERS

Sample	Diameter D(mm)	Length b (mm)	$g = g_1 = g_2$ mm^{-1}	gb	$\dfrac{gb}{\pi/2}$	n^a
I	2.0	8.33	0.188	1.57	~ 1	1.60
II	2.0	73.30	0.192	14.10	~ 9	1.60
III	2.2	498.00	0.173	86.40	~ 55	1.54

^{a} $n = \sqrt{\varepsilon(0)/\varepsilon_0}$ (refractive index)

samples used in this experiment have rather large diameters and small focusing constant g values compared with the fibers used for transmitting a single light beam, as shown in Table 5.4-I. The experimental apparatus is shown in Fig. 5.4-2. A He–Ne laser ($\lambda_0 = 0.633\ \mu$m) was used as a light source to avoid a chromatic aberration, thought to be one of the important problems of the lenslike medium, but this is not of interest in the present discussion. The image to be transmitted was the letter A, as shown in Fig. 5.4-3, printed on a transparent film. The transmitted image was focused directly on the photo plate.

The images transmitted by a system using sample II are shown in Figs. 5.4-4 and 5.4-5. In Fig. 5.4-4 the transmitted patterns are for various values of c,

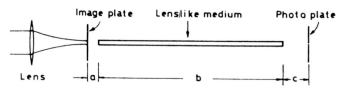

FIGURE 5.4-2. An experimental system of image transmission. (After Iga, Hata, Kato, and Fukuyo.[19])

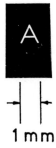

FIGURE 5.4-3. An image plate used for transmission experiments. (After Iga, Hata, Kato, and Fukuyo.[19])

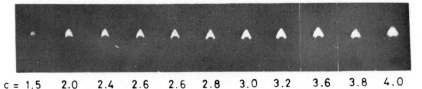

c = 1.5   2.0   2.4   2.6   2.6   2.8   3.0   3.2   3.6   3.8   4.0

FIGURE 5.4-4.   Patterns transmitted by an experimental system. The numbers are the distance c from the lens-like medium to the photo plate in millimeters. (After Iga, Hata, Kato, and Fukuyo.[19])

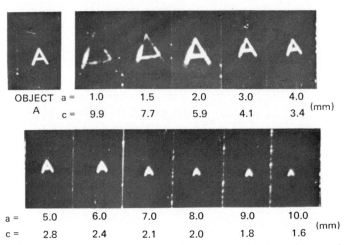

| OBJECT | a = | 1.0 | 1.5 | 2.0 | 3.0 | 4.0 | |
| A | c = | 9.9 | 7.7 | 5.9 | 4.1 | 3.4 | (mm) |

| a = | 5.0 | 6.0 | 7.0 | 8.0 | 9.0 | 10.0 | |
| c = | 2.8 | 2.4 | 2.1 | 2.0 | 1.8 | 1.6 | (mm) |

FIGURE 5.4-5.   Transmitted real images of the letter "A" (top left) for various a and c values, indicated in millimeters. (After Iga, Hata, Kato, and Fukuyo.[19])

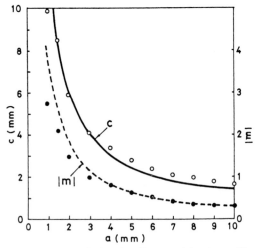

FIGURE 5.4-6.   Plotted data points for imaging distance c (○) and magnification |m| (●) versus a. Theoretical relations are also shown for c (solid line) and |m| (dashed line). (After Iga, Hata, Kato, and Fukuyo.[19])

when $a = 5$ mm. We see that the imaging condition is satisfied at $c = 2.4$ mm. The transmitted images for various values of a are shown in Fig. 5.4-5 for the imaging condition. The distance c and magnification m in this case are plotted in Fig. 5.4-6 as distance a is changed. Theoretical lines are also shown in this figure, obtained from Eqs. (5.4-22) and (5.4-24) by substituting the values in Table 5.4-I, and expressed by the equations

$$c = (0.207a + 10.6)/(a - 0.207), \qquad |m| = 3.25/(a - 0.207). \quad (5.4\text{-}30)$$

It must be added that image transmissions were successfully made on other samples, shown in Table 5.4-I. The results are not shown here because of space limitations.

5.5 FOURIER TRANSFORMS

A. FOURIER TRANSFORM CONDITION

We can obtain the Fourier transform condition from Eq. (5.4-14) by changing the order of integration and setting

$$\alpha = \alpha + \beta_i - \beta_i^2 \sec^2 g_i b/(\beta_i + \gamma),$$

which reduces in terms of Eqs. (5.4-5), (5.4-10), and (5.4-13) to

$$c = (\sqrt{\varepsilon(0)/\varepsilon_0}\, g_i \tan g_i b)^{-1}. \qquad (5.5\text{-}1)$$

The transformed field reduces to the phase-distorted Fourier transform, that is,

$$g(x, y) = -\frac{j\pi}{2\alpha} C \exp[-jB_1 x^2 - jB_2 y^2] F\left(\frac{2\beta_1\gamma \sec g_1 b}{\beta_1 + \gamma} x, \frac{2\beta_1\gamma \sec g_2 b}{\beta_2 + \gamma} y\right),$$

$$(5.5\text{-}2)$$

$$B_i = \frac{\beta_i\gamma}{\beta_i + \gamma} - \frac{1}{4\alpha}\left(\frac{\beta_i\gamma \sec g_i b}{\beta_i + \gamma}\right)^2, \qquad i = 1, 2, \qquad (5.5\text{-}3)$$

where C is defined by Eq. (5.4-15), and Eq. (5.4-19) has been used. From Eq. (5.5-1) we obtain distance c, which, for the Fourier transform, is determined only by the parameters of lens-like media, such as focusing constant g and length b.

B. EXPERIMENT VARIFICATION

The Fourier transform of a one-dimensional slit lying on the (x', y') plane shown in Fig. 5.4-1 or Fig. 5.4-2 was obtained by a system using sample II of

Table 5.4-I. Since the parameter gb is $\pi/2$ times an odd integer, the distance c for Fourier transform reduces to zero from Eq. (5.5-1). The transformed image of a slit 15 μm wide for various a's is shown in Fig. 5.5-1, where the transformed patterns coincide with the Fourier transform of the slit; the width of the peaks is about 190–200 μm for various a's, whereas the calculated value from Eq. (5.5-2) is 193 μm. It is noted from Fig. 5.5-1 that the width of the peaks is not affected by changing distance a; this is predicted by Eqs. (5.5-1) and (5.5-2).

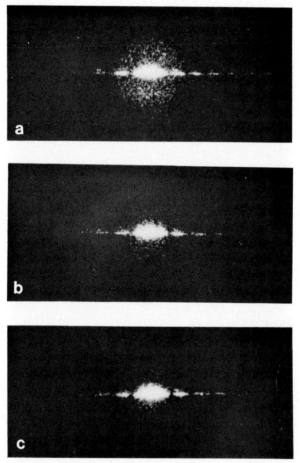

FIGURE 5.5-1.   Fourier transforms obtained by the experimental system at $c = 0$. The distance a of the slit from the medium was (a) 0 mm, (b) 1 mm, and (c) 2 mm. (After Iga, Hata, Kato, and Fukuyo.[19])

5.6 CONSIDERATION OF DISTORTION

In the imaging problem, β_{nm} itself plays an important role. If we neglect the third term in Eq. (5.2-5), β_{nm} can be approximated by a linear function of the mode number n. Thus, the spatial information ϕ_2 at the end of the GRIN medium (as shown in Fig. 5.4-1) with an arbitrary length is related to the input information ϕ_1 by an integral transform.[17] The following expression is for cylindrical coordinates:

$$\phi_2(r_2, \theta_2) = \frac{j}{\lambda b} \exp[-jk(0)b] \frac{gb}{\sin gb}$$

$$\times \int_0^\infty \int_0^{2\pi} r_1 \, dr_1 \, d\theta_1 \, \phi_1(r_1, \theta_1) K(r_1, \theta_1; r_2, \theta_2). \quad (5.6\text{-}1)$$

$$K(r_1, \theta_1; r_2, \theta_2) = \exp\left\{ -j\frac{k(0)g}{\tan gb}\left[r_1^2 - 2r_1 r_2 \frac{\cos(\theta_1 - \theta_2)}{\cos gb} + r_2^2 \right] \right\}. \quad (5.6\text{-}2)$$

This equation can be easily obtained by expanding $\phi_1(r_1, \theta_1)$ by the normal modes[12,13] and calculating the summation at the end of the medium. The necessary condition is that the propagation constant β_{nm} be a linear function of n and m, where m is the azimuthal mode number. The higher-order terms, therefore, correspond to the wave aberration. The measure of the degree of aberration is given by R, the ratio of the second and the third terms of Eq. (5.2-5).

If $N = 2n + m$ (principal mode number) and $1 - m^2$ is neglected in Eq. (5.2-5), R is reduced to

$$R \cong \tfrac{3}{4}[g/k(0)](h_4 - \tfrac{2}{3})(N + 1). \quad (5.6\text{-}3)$$

The maximum value of N is determined from the cutoff condition, that is, $\beta_{nm} = k(0)(1 - \Delta)$.

$$N_{\max} + 1 = \Delta[g/k(0)]^{-1}. \quad (5.6\text{-}4)$$

When $g/k(0) = 10^{-4}$ and $\Delta = 0.05$, $N_{\max} = 500$. If we guarantee $R < 0.1$, then $h_4 - \tfrac{2}{3}$ must be less than 2.

Since we know that the spatial information is transferred in the free space in terms of a 2-D convolution[12] with the help of Eq. (5.7-1) and the convolution, the transformation relationship can be derived[17] for the optical system shown in Fig. (5.4-1).

A coherent imaging condition, including phase shifts, and a Fourier transform condition were discussed in Section 5.5. In Reference 20 the integral transform of a free space was used approximately, but here we have described the exact shape of the integral for a DI medium.

5.7 LIMIT OF SPATIAL FREQUENCY

A. IMPULSE RESPONSE AND SYSTEM FUNCTION

The limits of the integration in Eq. (5.2-3) are assumed to extend to infinity. In actuality, the spatial frequency that can be transmitted by the system may be limited by the finiteness of the diameter of the lenslike medium. In this section, we shall discuss this problem on the basis of a two-dimensional impulse response and system function.

First, we regard a system like that in Fig. 5.4-1 as a two-dimensional system. Let us obtain the response $h(x, y; x_0, y_0)$ for the two-dimensional spatial impulse

$$\delta(x' - x_0)\delta(y' - y_0),$$

where $\delta(x' - x_0)$ is the Dirac delta function with peak at $x' = x_0$, since we know that the response $g(x, y)$ is associated with any function $f(x, y)$ from the following relation:

$$\phi_2(r_2, \theta_2) = \frac{j}{\lambda b} \exp[-jk(0)b] \frac{gb}{\sin gb}$$

$$\times \int_0^\infty \int_0^{2\pi} r_1 \, dr_1 \, d\theta_1 \, \phi_1(r_1, \theta_1) K(r_1, \theta_1; r_2, \theta_2). \quad (5.7-1)$$

The impulse response is obtained from Eq. (5.4-13), when the limits of the integration of x_1 and y_1 approach infinity and the imaging condition as in Eq. (5.4-21) is fulfilled. It can be written as

$$h(x, y; x_0, y_0) = C(\pi/\alpha)^2 \, m_1 \exp[-j\beta_1\gamma/(\beta_1 + \gamma)x^2]$$

$$\times \exp(-j\alpha x_0^2) \, \delta(x - m_1 x_0)$$

$$\times \exp[-j\beta_2\gamma/(\beta_2 + \gamma)y^2]$$

$$\times \exp(-j\alpha y_0^2) \, \delta(y - m_2 y_0), \quad (5.7-2)$$

where Eq. (5.4-24) is used.

If we define the system function $H(u, v)$ as a Fourier transform of $h(x, y; x_0, y_0)$ in terms of Eq. (5.4-19), and normalize $H(u, v)$ by $H(0, 0)$, we obtain from Eq. (5.8-2):

$$\bar{H}(u, v) = H(u, v)/H(0, 0). \quad (5.7-3)$$

B. FINITE DIAMETER

When the lenslike medium has a finite diameter or the aperture is used to avoid aberration, the range of integration in Eq. (5.4-14) must be modified. If we consider the effective rectangular aperture of $2A_e \times 2B_e$ just in front of the lenslike medium, the limits of the integration may be from $-A_e$ to A_e in the x direction and from $-B_e$ to B_e in the y direction. In this case; the impulse response reduces to the following form:

$$h(x, y; x_0, y_0) = C \exp[j\beta_1\gamma/(\beta_1 + \gamma)x^2 - j\alpha x_0^2] \frac{\sin(2\alpha A_e/m_1)(x - m_1 x_0)}{(2\alpha/m_1)(x - m_1 x_0)}$$

$$\times \exp[j\beta_2\gamma(\beta_2 + \gamma)/(\beta_2 + \gamma)^2 y^2 - j\alpha y_0^2] \frac{\sin(2\alpha Ae/m_2)(y - m_2 y_0)}{(2\alpha m_2)(y - m_2 y_0)}$$

$$(5.7\text{-}4)$$

The system function associated with Eq. (5.8-4) is obtained from Eq. (5.4-19) and written as

$$|\bar{H}(u, v)| = p_{2\alpha A_e/m_1}(u)p_{2\alpha B_e/m_2}(v), \tag{5.7-5}$$

where $p_D(u)$ is the step function as defined by the equation

$$p_D(u) = \begin{cases} 1, & |u| \le D, \\ 0, & |u| > D. \end{cases} \tag{5.7-6}$$

The result shown in Eq. (5.7-5) will be generalized; the cutoff spatial frequency $u_c/2\pi$ is determined by the effective aperture width $2A_e$ and expressed by the equation

$$\frac{u_c}{2\pi} = \frac{\alpha A_e}{m\pi} = \frac{k_0}{2\pi m}\frac{A_e}{a}. \tag{5.7-7}$$

C. Effective Diameter

Suppose the impulse point source exists at point P on the (x', y') plane with $x' = x_0 (>0)$ and $y' = 0$, as shown in Fig. 5.7-1. In addition, we assume that $a = c$, $g_1 b = g_2 b = 2\pi N + 3\pi/2$, with N a positive integer, and the diameter of the lenslike medium is $2A$. This means that the system transmits the real

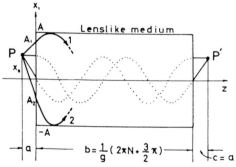

FIGURE 5.7-1. Effective diameter of a lens-like medium. The rays 1 and 2 are those which pass the edge of the medium. The rays indicated by dotted lines are representative of ones that make the image P' of the point source P. (After Iga, Hata, Kato, and Fukuyo.[19])

image P' with multiplication $m = +1$. We shall now examine the effective diameter $2A_e$ and the cutoff spatial frequency. According to ray optics, the maximum ray deviation from the center axis must be smaller than the edge of the lenslike medium. The amplitude of ray 1 or 2 in Fig. 5.7-1, which is incident into the lenslike medium from the point source P at $x_1 = A_1$, $y_1 = 0$ or $x_1 = A_2$, $y_1 = 0$ and passes the edge of the lenslike medium, is obtained as follows:

$$\left[A_i^2 + \left(\frac{x_0 - A_i}{nga} \right)^2 \right]^{1/2}, \qquad i = 1 \quad (\text{or } 2),$$

where n is the refractive index equal to $\sqrt{\epsilon(0)/\epsilon_o}$ at the center of the medium. This amplitude must be equal to A, leading us to the effective edge A_1 and A_2:

$$2A_{1,2} = A[x_0/A \pm \sqrt{2 - x_0^2/A^2}], \qquad (5.7\text{-}8)$$

where the imaging condition $nga = 1$, taken from Eq. (5.4-22), is used. The effective width $2A_e$ is, therefore, obtained by subtracting A_2 from A_1 according to Eq. (5.7-8) and given by

$$2A_e = A_1 - A_2 = A\sqrt{2 - x_0^2/A^2}. \qquad (5.7\text{-}9)$$

The cutoff frequency $u_c/2$ reduces, from Eqs. (5.7-7) and (5.7-9) with $m = 1$ and $a = 1/ng$, taken from the imaging condition as in Eq. (5.4-22), to the following:

$$\frac{u_c}{2\pi} = \frac{A_e}{\lambda_0 a} = \frac{ngA}{2\lambda_0} \sqrt{\frac{2 - x_0^2}{A^2}} \qquad (5.7\text{-}10)$$

The numerical examples of $u_c/2\pi$ for $\lambda_0 = 0.63$ μm and $n = 1.6$ are shown in Fig. 5.7-2. If we put $g = 0.2$ mm^{-1} and $A = 1$ mm, the cutoff frequency $u_c/2\pi$ associated with the center axis ($x_0 = 0$) is read as 356 mm^{-1}, and it is reduced by one-half for $x_0 = 1.3A$.

D. Lossy Medium

Let us next examine the impulse response and the system function when the medium effects a loss, as mentioned in Section 5.4. The impulse response is obtained according to the integral in Eq. (5.4-7) and expressed by the equation

$$h(x, y; x_0, y_0) \propto \exp[-\tfrac{1}{2}(x - m_1 x_0)^2/\Delta_1^2 - \tfrac{1}{2}(y - m_2 x_0)^2/\Delta_2^2], \qquad (5.7\text{-}11)$$

where

$$\Delta_i = w_{0i}\sqrt{[1 + 1/(4w_{0i}^4 \gamma^2)]} \tanh \Gamma_i b, \qquad i = 1, 2. \qquad (5.7\text{-}12)$$

The system function $\bar{H}(u, v)$ therefore, is calculated in terms of Eq. (5.7-11) as

$$u_c = \frac{1}{\Delta_1} = \frac{1}{w_{0i}\sqrt{[1 + 1/(4w_{0i}^4 \gamma^2)]} \tanh \Gamma_i b}, \qquad (5.7\text{-}13)$$

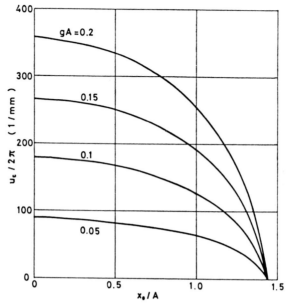

FIGURE 5.7-2. Cut-off frequency $u_c/2\pi$ versus the normalized distance x_0/A of the point source from the center axis. (After Iga, Hata, Kato, and Fukuyo.[19])

where

$$|\bar{H}(u, v)| = \exp[-\tfrac{1}{2}u^2/u_c^2 - \tfrac{1}{2}v^2/v_c^2].\qquad(5.7\text{-}14)$$

The cutoff frequency u_c means the frequency at which the intensity frequency response $|\bar{H}(u, v)|^2$ reduces to $1/e$ of $|\bar{H}(0, 0)|^2$. The experimental data for the cutoff frequency were obtained by Hamasaki et al.[2] and compared with this theoretical result.

5.8 DIFFRACTION AND FOCUSING LIMIT

Focusing a light beam into a small spot is an important application of the DI lens. Let us consider here the limitation of the focused spot and the design of a focuser.

One limit of the focused spot size is determined by diffraction. We shall consider the $\tfrac{1}{4}$-pitch DI lens, that is, $gb = \pi/2$; the incident light is assumed to be a plane wave. We then have

$$\phi_1(r_1) = \begin{cases} 1, & 0 < r_1 < A, \\ 0, & r_1 > A, \end{cases}\qquad(5.8\text{-}1)$$

in Eq. (5.7-1).

The integral can be easily performed and reduces to

$$\phi_2(r_2) = C \frac{A^2}{2} \frac{2J_1[k(0)gr_2A]}{k(0)gr_2A}. \tag{5.8-2}$$

As the half-width at half-maximum of $\{2J_1(x)/x\}^2$ is approximately $\frac{3}{2}$, the spot diameter due to diffraction is

$$\Delta D_d \cong \frac{3}{k(0)gA} \cong \frac{1}{2} \frac{\lambda}{NA^*}. \tag{5.8-3}$$

Here $NA^* = n(0)gA$, which corresponds to the numerical aperture.

The second limit comes from the aberration as mentioned above. The intensity of the spot may be visualized in terms of a ray tracing, as shown in Fig. 3.3-2. The horizontal axis r' is normalized by $\frac{3}{4}|h_4 - \frac{2}{3}|(gA)^2(gb)$. The half-maximum of the spot can be obtained from the figure, and the diameter D_a is given by

$$\Delta D_a = (0.084/g)|h_4 - \frac{2}{3}|(gb)(gA)^3. \tag{5.8-4}$$

The spot diameter D_a due to the aberration to the fourth order is proportional to $(gA)^3$. By combining Eqs. (5.4-17) and (5.4-18), we obtain the limit of the focused spot as shown in Fig. 5.8-1.

On the other hand, when we focus light by a thin lens, the diffraction limit is calculated as follows. Let us assume that the coordinate z as an optical axis and

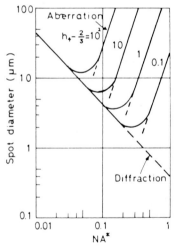

FIGURE 5.8-1.   A limit of the focused spot by a $\frac{1}{4}$-pitch lens, where $n(0) = 1.5$, $g = 0.17$ mm^{-1}, and $\lambda = 0.8$ μm. (After Iga.[22])

lens is located at $z = 0$ as shown in Fig. 5.8-2. The light wave through the lens becomes a converging wave and is expressed by

$$f(r', \theta', 0) = \exp[jk(r'^2/2f_0)] \qquad (5.8\text{-}5)$$

at $z = 0$ where the prime means the coordinate $z = 0$. Light field at any desired position can be calculated using a Fresnel integral in free space [Eq. (5.3-15)]. Our present interest is to obtain a field distribution of the focused spot in the focal plane which is perpendicular to the z axis at $z = f_0$. Substitution of Eq. (5.8-5) into Eq. (5.3-15) results in

$$f(r, \theta, f_0) = \frac{j}{\lambda f_0} \exp[-jk(f_0 + (r^2/2f_0))].$$

$$\times \int_0^A \int_0^{2\pi} d\theta' \, r \, dr' \exp\left[\frac{jk}{f_0} rr' \cos(\theta - \theta')\right], \qquad (5.8\text{-}6)$$

where A is the upper limit of integral with respect to r'.

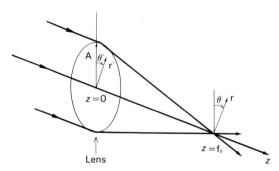

FIGURE 5.8-2. Coordinate system for calculating diffraction pattern of converging beam.

The phase change with respect to r' is canceled by the phase distribution of converging wave expressed by Eq. (5.8-6). The integration of Eq. (5.8-6) can be done first with respect to θ' and then we have

$$f(r, \theta, f_0) = j\frac{2\pi}{\lambda f_0} \exp[-jk(f_0 + (r^2/2f_0))] \int_0^A r' \, dr' \times J_0\left(\frac{2\pi}{\lambda} \frac{rr'}{f_0}\right), \qquad (5.8\text{-}7)$$

where the integral expression of Bessel function has been used; i.e., for the Bessel function of order θ is expressed by

$$J_0(u) = \frac{1}{2\pi} \int_\delta^{2\pi+\delta} \exp(-ju \sin\theta') \, d\theta', \qquad (5.8\text{-}8)$$

where δ is an arbitrary constant. We can replace θ' by $(\theta + \theta') + \frac{1}{2}\pi$ and δ by $\theta + \frac{1}{2}\pi$,

$$J_0(u) = \frac{1}{2\pi} \int_0^{2\pi} \exp[ju\cos(\theta - \theta')]\, d\theta'. \qquad (5.8\text{-}9)$$

Using Eq. (5.8-9) the integration from 0 to lens radius A with respect to r' reduces to

$$f(r, \theta, f_0) = j\frac{2\pi A^2}{\lambda f_0} \exp\left[-jk\left(f_0 + \frac{r^2}{2f_0}\right)\right]\frac{J_1(\rho)}{\rho}, \qquad (5.8\text{-}10)$$

where $\rho = 2\pi Ar/\lambda f_0$, and we have used the relation

$$\int_0^A r' J_0(\alpha r')\, dr = A/\alpha J_1(\alpha A). \qquad (5.8\text{-}11)$$

The intensity distribution $I(\rho)$ is obtained by

$$I(\rho) = |f(r, \theta, f_0)|^2 = \frac{\pi^2 A^4}{\lambda^2 f_0^2}\left(\frac{2J_1(\rho)}{\rho}\right)^2. \qquad (5.8\text{-}12)$$

The intensity distribution $I(\rho)$ is the maximum at $\rho = 0$ and oscillately decreases with respect to ρ. In Fig. 5.8-3, $I(\rho)/I(0)$ is shown. Since $\rho = 3.8$ is the first zero of $I(\rho)$, spot size D_S is calculated as

$$D_S \simeq 1.22\,\frac{f_0\lambda}{A} = 1.22\,\frac{\lambda}{\text{NA}}, \qquad (5.8\text{-}13)$$

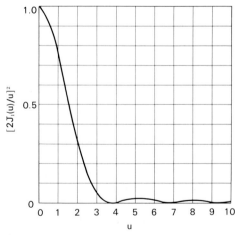

FIGURE 5.8-3. $[2J_1(u)/u]^2$

when we define the spot size as the diameter of the first zero of the Airly disks. The spot size is sometimes defined by the diameter where its intensity become half maximum. In this case the spot size is given by

$$D_S \simeq 0.5 \times \frac{f_0 \lambda}{A} = 0.5 \times \frac{\lambda}{NA}, \qquad (5.8\text{-}14)$$

since $\rho \simeq \frac{3}{2}$ when $[2J_1(\rho)/\rho]^2 = \frac{1}{2}$.

5.9 STEP-INDEX IMAGING

A light ray that incidents with a certain angle θ_{in} on the surface of a step-index optical fiber with uniform core index n_1 will excite a corresponding waveguide mode.[23] The light ray is associated with the mode, that is, the angle θ_N of the ray with respect to the optical axis is expressed by

$$\theta_N = n_1 \sin \theta_N \cong (\pi/2k_0 a)(N + 1) = (\lambda/4a)(N + 1), \qquad (5.9\text{-}1)$$

where N is the mode number, a the core radius, and $k_0 = 2\pi/\lambda$. The angle θ_{in} is related to θ_N by

$$\sin \theta_{in} = n_1 \sin \theta_N, \qquad (5.9\text{-}2)$$

or $\theta_{in} \cong n_1 \theta_N$ when θ_{in} is not so large.

After propagating along the fiber core, the guided light will exit from the other end of the fiber. For a step-index optical fiber, the output light forms a cone-shaped beam, due to skew rays, with a solid angle θ_{out}, which corresponds to the input incidence angle. This beam will make a ring pattern on the observation plane. The angle θ_{out} may also be expressed by

$$\theta_{out} \cong n_1 \theta_1. \qquad (5.9\text{-}3)$$

This phenomenon shows that an image can be transmitted by the optical fiber, that is, there is a relation between spatial information of the output pattern and the angular position of the light source, unless mode mixing is not appreciable. Since the cone-shaped beam that spreads out from the end of the fiber is due to skew rays, it is essential to eliminate the output light, except in a suitable direction associated with the position of the point source. If we use a pair of slits in front of and behind the fiber to do this, we can obtain an actual point image instead of a ring pattern, as shown in Fig. 5.9-1. For the purpose of transmitting two-dimensional images, we may rotate the input and output slits simultaneously with a definite angular relationship. This is the method we shall propose here for reconstructing the image.

The experimental arrangement is shown in Fig. 5.9-1. Light rays from an object are focused by a lens on the surface of the input end of the fiber. The angular position of a given point source on the object plane determines the

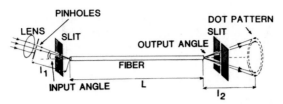

FIGURE 5.9-1.   Experimental configuration for a step-index optical fiber. (After Batubara, Iga, and Oikawa.[23])

input-angle incidence to the fiber surface. We used a sample of a fiber with the specifications listed in Table 5.9-1. To investigate the relationship between the output and input angles, we measured the diameter of the ring patterns for various incident angles of input light. Assuming the output angles are one-half of the solid angles of the output light, we plotted those values versus the corresponding input angles, as shown in Fig. 5.9-2. The distance l_1 between the input end of the fiber and the object is 14.5 cm, and the focal length of the lens is 15.5 cm. The fiber used was 30.9 cm long, and the other fibers exhibited similar results. There was some variation of the output angle for a given input

TABLE 5.9-I

SPECIFICATIONS OF FIBER SAMPLES

Fiber	Core diameter (μm)	Cladding diameter (μm)	Δ (%)	Length (cm)
1	102	150	1.10	30.9
2	100	150	(NA = 0.28)	22.8
3	50	125	0.76	23.6

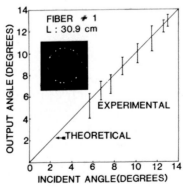

FIGURE 5.9-2.   Relationship between output and input angles for a step-index optical fiber. The ring pattern is shown in the inset, where $L = 30.9$ cm. (After Batubara, Iga, and Oikawa.[23])

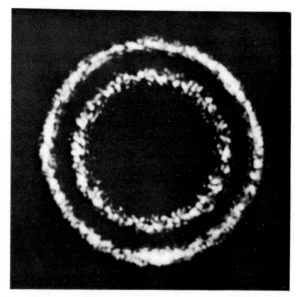

FIGURE 5.9-3. Ring pattern formed by a sample fiber. (After Batubara, Iga, Oikawa.[23])

angle. This is considered to be due to the excitation condition, and the spread might be reduced by preparing more carefully the surface of the input and output fiber ends and optimizing the focal length of the lens. A ring pattern of the output light of two point sources is shown in Fig. 5.9-3.

To obtain the image of a pair of dot patterns, we used a slit placed next to the output end of the fiber. The object used consisted of a plane with an array of pinholes illuminated by a light source from a He–Ne laser. By using a slit placed behind the output end of the fiber, we observed a pair of dot patterns of two point sources. Figure 5.9-4 shows the pattern for two point sources, with a

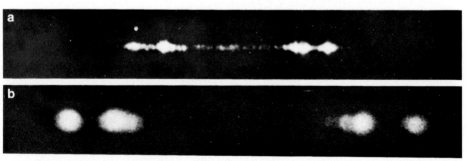

FIGURE 5.9-4. Dot patterns for two point sources having a separation of (a) 4.5 mm and (b) 8.0 mm. (After Batubara, Iga, and Oikawa.[23])

separation of 4.5 mm. When we look at one of the images (mirror symmetry), we can say that it is the image of the object transmitted through a single piece of fiber. Of course, the image is not formed simultaneously because part of the image is transmitted with a different group velocity.

5.10 CONCLUDING REMARKS

In this chapter the image transmissions through a DI medium were treated on the basis of wave optics. The results obtained are summarized as follows:

(1) The general form of image transforms by a DI medium is derived in terms of the integral transform. Because this treatment conserves the phase relation, the results can be applied not only to incoherent imaging but also to coherent imaging, such as the transmission of a reference wave or signal wave in a holography using a DI medium.[24]

(2) The imaging condition is derived from the transform law and compared with experimental results.

(3) It is shown both theoretically and experimentally that the DI medium is applied to the two-dimensional Fourier transform.

(4) The limitation of the spatial frequency is obtained. The cutoff spatial frequency $u_c/2\pi$ associated with the off-axis image whose center exists at a distance x_0 from the center axis is expressed by the equation

$$u_c/2\pi = (ngA/2\lambda_0)\sqrt{2 - x^2/A^2},$$

where $n, g, 2A$, and λ_0 are the refractive index, focusing constant, diameter, and wavelength, respectively.

(5) The cutoff frequency in the case of an absorptive DI medium is found to be proportional to the inverse of $\sqrt{\tanh \Gamma b}$, where Γ and b are the attenuation constant of Hermite–Gaussian modes and the length of the medium, respectively.

(6) Step-index imaging is introduced as an example of pulse-modulated transmission of images.

REFERENCES

1. N. S. Kapany, *Fiber Optics*. Academic Press, New York, 1967.
2. Y. Suematsu, *Kogaku* 1, 175 (1972) [in Japanese].
3. E. A. J. Marcatili, *Bell Syst. Tech. J.* **43**, 2887 (1964).
4. D. Marcuse, *Light Transmission Optics*, Van Nostrand–Reinhold, New York, 1972.
5. D. W. Berreman, *Bell Syst. Tech. J.* **43**, 1469 (1964).
6. Y. Suematsu, K. Iga, and S. Ito, *IEEE Trans. Microwave Theory and Tech.* **MTT-14**, 657 (1966).
7. T. Uchida, M. Furukawa, I. Kitano, K. Koizumi, and H. Matsumura, *IEEE J. Quantum Electron.* **QE-5**, 331 (1969).

8. F. P. Kapron, D. B. Keck and R. D. Maurer, *Trunk Communications by Guided Waves*, *IEEE Conf. Pub.* **148**, (1970).
9. Y. Aoki and M. Suzuki, *IEEE Trans. Microwave Theory Tech.* **MTT-15**, 1 (1967).
10. T. Uchida, M. Furukawa, I. Kitano, K. Koizumi, and H. Matsumura, *IEEE J. Quantum Electron.* **QE-6**, 606 (1970).
11. E. N. Leith and J. Upatnieks *J. Opt. Soc. Am.* **52**, 1123 (1962).
12. A. Papoulis, *Systems and Transforms with Applications in Optics*. McGraw–Hill, New York, 1968.
13. K. Iga and S. Hata, *IECE Jpn. Rep. Tech. Group Quant. Elect.* **QE71-56** (1972).
14. Y. Suematsu and K. Iga, *J. Inst. Elect. Commum. Eng. Jpn.* **49**, 59 (1966).
15. Y. Suematsu and K. Furuya, *Trans. IECE Jpn.* **54-B**, 325 (1971) [in Japanese].
16. S. Kawakami and J. Nishizawa, *IEEE Trans. Microwave Theory Tech.* **MTT-16**, 814 (1968).
17. Y. Suematsu, *IECE Jpn. Rep. Tech. Group Microwave*, January (1967).
    Y. Suematsu and H. Fukinuki, *J. IECE Jpn.* **48**, 1684 (1965).
18. Y. Suematsu, T. Shimizu, and T. Kitano, *Trans. IECE Jpn.* **53-B**, 727 (1970) [in Japanese].
19. K. Iga, S. Hata, Y. Kato, and H. Fukuyo, *J. Appl. Phys.* **13**, 79 (1974).
20. A. Yariv, *J. Opt. Soc. Am.* **66**, 301 (1976).
21. J. Hamasaki and K. Maeda, *IECE Japan Natl Conv. Rec.* (1001), 1012 (1973).
22. K. Iga, *Appl. Opt.* **19**, 1041 (1980).
23. J. E. Batubara, K. Iga, and M. Oikawa, *Jpn. J. Appl. Phys.* **21**, L749 (1982).
24. H. Fukuyo, K. Iga, and Y. Kato, *IECE Japan Natl Conv. Res.* (991), 994 (1973).

CHAPTER 6

Beam Optics

A Gaussian beam does not change distribution when propagating through free space and in a DI medium. Mathematically, it is invariant in terms of Fourier and Fresnel transforms. The Gaussian mode is thus very important in light-wave transmission, especially in a lens guide system. The results discussed in this chapter will be relevant to the treatment of stacked planar optics in Chapter 10.

6.1 GAUSSIAN BEAM

The Fresnel–Kirchhoff integral for transformation of images through free space and a DI medium has been discussed. This chapter will describe how a Gaussian beam changes form when propagating in free space.

We shall assume a Gaussian beam at $z = 0$ given by

$$f(x', y', 0) = E_0 \exp\left(-\frac{1}{2} \frac{x'^2 + y'^2}{s^2} \right). \tag{6.1-1}$$

Positional change of the beam can be calculated by Eq. (5.3-14), and expressed as

$$f(x, y, z) = E_0 \left[\exp(-jkz)\right](s/w) \exp\left[-\tfrac{1}{2} P(x^2 + y^2) + j\varphi \right]. \tag{6.1-2}$$

Here the spotsize w and radius R of the phase front are given by

$$w = s\sqrt{1 + (z/ks^2)^2}, \tag{6.1-3}$$

$$R = z[+ (ks^2/z)^2]. \tag{6.1-4}$$

Parameters P and φ are defined by

$$P = (1/w^2) + j(k/R), \tag{6.1-5}$$

$$\varphi = \tan^{-1}(z/ks^2). \tag{6.1-6}$$

From Eq. (6.1-2) it can be seen that the transformed beam is still Gaussian, although there are spotsize and phase front changes. It is clear that R expresses the phase front if consideration is made of the phase condition

$$kz + (k/2R)r^2 = \text{const.}$$

With this equation, the functional dependence of phase front $z = -\tfrac{1}{2}Rr^2$ can be reduced. When R is positive, the phase front is convex, as seen from $z = +\infty$.

Let us next examine parameter z/ks^2 that appeared in Eqs. (6.1-3) and (6.1-4). When this parameter is rewritten as

$$z/ks^2 = (1/2\pi)(s^2/\lambda z)^{-1}, \tag{6.1-7}$$

and the Fresnel number N is defined as

$$N = s^2/\lambda z, \tag{6.1-8}$$

the Fresnel number is a function of wavelength, distance, and spotsize and expresses normalized distance. Regions can be characterized according to N such that

$$N \ll 1 \quad \text{(Fraunhofer region)},$$

$$N \gtrsim 1 \quad \text{(Fresnel region)}.$$

When the point of observation is located at a point some distance from the origin ($N \ll 1$), the spotsize w can be approximated from Eq. (6.1-4) as

$$w \simeq z/ks. \tag{6.1-9}$$

The spreading angle of the beam is, therefore,

$$2\,\Delta\theta = 2w/z = 0.64\lambda/2s. \tag{6.1-10}$$

This is analogous to the spreading angle of a main lobe of a diffracted plane wave from a circular aperture, given by

$$2\,\Delta\theta = 1.22\lambda/D. \tag{6.1-11}$$

6.2 WAVEFORM MATRIX[1]

Figure 6.2-1 presents waveform coefficients P_0, P_1, and P_2, at $z = 0, z_1$, and z_2, respectively. If the spotsizes and curvature radii of the wavefront are given by s, w_1 and w_2, and ∞, R_1, and R_2, respectively, the coefficients can be expressed as

$$P_0 = 1/s^2, \tag{6.2-1}$$

$$P_1 = (1/w_1^2) + (jk/R_1) \tag{6.2-2}$$

$$P_2 = (1/w_2^2) + (jk/R_2) \tag{6.2-3}$$

From Eqs. (6.2-1)–(6.2-3),

$$1/P_0 = (1/P_1) + (jz_1/k), \tag{6.2-4}$$

$$1/P_0 = (1/P_2) + (jz_2/k). \tag{6.2-5}$$

When P_0 is eliminated, the relationship between P_1 and P_2 is reduced to

$$P_1 = P_2/[1 + (j/k)(z_2 - z_1)P_2]. \tag{6.2-6}$$

This is a special case of the linear transform

$$P_1 = (AP_2 + B)/(CP_2 + D). \tag{6.2-7}$$

It is very convenient to utilize the matrix form[1]

$$\tilde{F} = \begin{bmatrix} A & B \\ C & D \end{bmatrix} \tag{6.2-8}$$

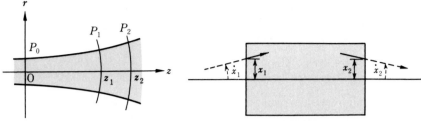

FIGURE 6.2-1. Transformation of waveform coefficients.

FIGURE 6.2-2. Relationship of ray positions and ray slopes.

TABLE 6.2-1

WAVEFORM MATRICES FOR VARIOUS OPTICAL SYSTEMS

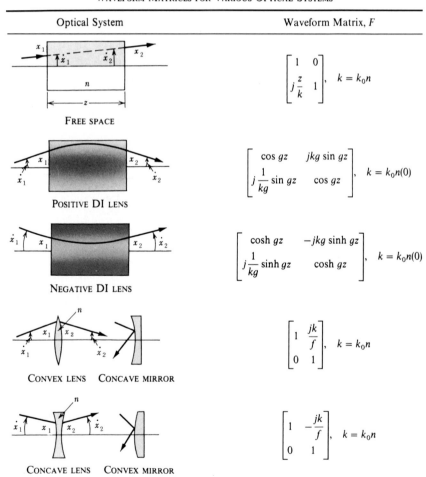

Optical System	Waveform Matrix, F
FREE SPACE	$\begin{bmatrix} 1 & 0 \\ j\dfrac{z}{k} & 1 \end{bmatrix}$, $k = k_0 n$
POSITIVE DI LENS	$\begin{bmatrix} \cos gz & jkg\,\sin gz \\ j\dfrac{1}{kg}\sin gz & \cos gz \end{bmatrix}$, $k = k_0 n(0)$
NEGATIVE DI LENS	$\begin{bmatrix} \cosh gz & -jkg\,\sinh gz \\ j\dfrac{1}{kg}\sinh gz & \cosh gz \end{bmatrix}$, $k = k_0 n(0)$
CONVEX LENS CONCAVE MIRROR	$\begin{bmatrix} 1 & \dfrac{jk}{f} \\ 0 & 1 \end{bmatrix}$, $k = k_0 n$
CONCAVE LENS CONVEX MIRROR	$\begin{bmatrix} 1 & -\dfrac{jk}{f} \\ 0 & 1 \end{bmatrix}$, $k = k_0 n$

to calculate the transform for a system composed of many tandem components. It is then possible to realize a total $\tilde{F}$ matrix with the product of the matrices, expressed as

$$\tilde{F} = \tilde{F}_1 \times \tilde{F}_2 \times \tilde{F}_3 \times \cdots . \tag{6.2-9}$$

Table 6.2-1 presents a tabulation of the waveform matrices associated with some optical components. It is not difficult to obtain these matrix forms by calculating the change of a Gaussian beam when it passes through these optical components (refer to Ref. 1 for the reduction of matrices). Kogelnik also proposed a matrix form for the same purpose,[2] but it is somewhat different from the definition introduced here.

6.3 RAY MATRIX

Figure 6.2-2 shows that the ray position x_1 and ray slope $\dot{x}_1$ at the incident position are related to x_2 and $\dot{x}_2$ by the same matrix representation; that is,

$$\begin{bmatrix} jk\dot{x}_1 \\ x_1 \end{bmatrix} = \begin{bmatrix} A & B \\ C & D \end{bmatrix} \begin{bmatrix} jk\dot{x}_2 \\ x_2 \end{bmatrix}. \tag{6.3-1}$$

The propagation constant k is included in Eq. (6.3-1) to make it possible to treat a tandem connection of optical components having different refractive indices.

REFERENCES

1. H. Fukinuki and Y. Suematsu, *J. IECE Jpn.* **48**, 1684 (1965).
2. H. Kogelnik, *Appl. Opt.* **4**, 1562 (1965).

6.4 IMAGING CONDITION (added in proof)

We shall apply the ray matrix method to the derivation of imaging condition. Let the ray matrix of the optical device $\tilde{F} = (A/C \, B/D)$. The total matrix is calculated by the product of the ray matrices for free space by referring Eq. (6.41).

$$\tilde{F} = \begin{pmatrix} 1 & 0 \\ j\dfrac{a}{k_0} & 1 \end{pmatrix} \begin{pmatrix} A & B \\ C & D \end{pmatrix} \begin{pmatrix} 1 & 0 \\ j\dfrac{c}{k_0} & 1 \end{pmatrix} = \begin{pmatrix} A & B \\ C + j\dfrac{a}{k_0}A & B \cdot j\dfrac{a}{k_0} + D \end{pmatrix} \begin{pmatrix} 1 & 0 \\ j\dfrac{c}{k_0} & 1 \end{pmatrix}$$

$$= \begin{pmatrix} A + j\dfrac{c}{k_0}B & B \\ C + j\dfrac{a}{k_0}A + \left(B \cdot j\dfrac{a}{k_0} + D\right) \cdot j\dfrac{c}{k_0} & B \cdot j\dfrac{a}{k_0} + D \end{pmatrix} \equiv \begin{pmatrix} A_T & B_T \\ C_T & D_T \end{pmatrix}. \tag{6.4-1}$$

If we assume that the image is formed by this optical system, the position of the ray x_2 on the image plane remains constant and proportional to x_1 even if we change the slope of rays coming out from the object. We express this as $x_2 = mx_1$ where m is the magnification of the image.

From Eq. (6.4-1) we get

$$\begin{pmatrix} jk_0\dot{x}_1 \\ x_1 \end{pmatrix} = \begin{pmatrix} A_T & B_T \\ C_T & D_T \end{pmatrix} \begin{pmatrix} jk_0\dot{x}_2 \\ x_2 \end{pmatrix} \tag{6.4-2}$$

$$jk_0\dot{x}_1 = A_T(jk_0\dot{x}_2) + B_T x_2, \qquad x_1 = C_T(jk_0\dot{x}_2) + D_T x_2. \tag{6.4-3}$$

If the aforementioned statement is true for imaging condition, we obtain

$$C_T = C + j\frac{a}{k_0}A + \left(B \cdot j\frac{a}{k_0} + D\right) \cdot j\frac{c}{k_0} = 0 \tag{6.4-4}$$

$$m = D_T^{-1} = \frac{1}{D + j(a/k_0) \cdot B}. \tag{6.4-5}$$

First, let us consider the simplest case that the focusing device is a convex lens with focal length f;

$$\begin{pmatrix} A & B \\ C & D \end{pmatrix} = \begin{pmatrix} 1 & j\dfrac{k_0}{f} \\ 0 & 1 \end{pmatrix}. \tag{6.4-6}$$

The imaging condition reduces to be $1/a + 1/c = 1/f$. This is a well-known image forming condition. The magnification m is

$$m = \frac{1}{1 - a/f} = -\frac{c}{a}. \tag{6.4-7}$$

The minus sign means the image is inverted. The second example is DI imaging:

$$\begin{pmatrix} A & B \\ C & D \end{pmatrix} = \begin{pmatrix} \cos gb & jkg \sin gb \\ j\dfrac{1}{kg}\sin gb & \cos gb \end{pmatrix} \tag{6.4-8}$$

The C_T component is then

$$C_T = \left(j\frac{1}{kg} - jkg\frac{ac}{k_0}\right)\sin gb + \left(j\frac{a}{k_0} + j\frac{c}{k_0}\right)\cos gb$$

$$= j\left\{\left(\frac{1}{kg} - \frac{kg}{k_0^2}ac\right)\sin gb + \frac{1}{k_0}(a + c)\cos gb\right\} = 0 \tag{6.4-9}$$

$$\tan gb = -\frac{a + c}{(1/kg) - (kg/k_0^2)ac} \cdot \frac{1}{k_0} = -\frac{a + c}{1/n(0)g - n(0)gac} = -\frac{n(0)g(a + c)}{1 - n^2(0)g^2 ac}. \tag{6.4-10}$$

The magnification m is given by

$$m = \left[\cos gb + j\frac{a}{k_0} \cdot jkg \sin gb\right]^{-1} = [\cos gb - n(0)ga \sin gb]^{-1}. \tag{6.4-11}$$

For some special cases, the associated magnification is listed in Table 6.4-I.

TABLE 6.4-I

gb	$\tan gb$	Imaging condition	m
$\pi/2$	$+\infty$	$ac = \dfrac{1}{n^2(0)g^2}$	$-\dfrac{1}{n(0)ga}$
π	0	$a + c = 0$	-1
$3\pi/2$	$-\infty$	$ac = \dfrac{1}{n^2(0)g^2}$	$+\dfrac{1}{n(0)ga}$
2π	0	$a + c = 0$	$+1$

CHAPTER 7

Distributed-Index Formation

7.1 OPENING REMARKS

DI media[1] having an index distribution of the form shown by Eq. (2.2-2) evidence imaging properties.[2] The focusing constant g can be approximated by

$$g \simeq \sqrt{2\Delta}/a, \qquad (7.1\text{-}1)$$

where $\Delta = [n(0) - n(a)]/n(0)$ and a is the core radius. Imaging properties are realized because light rays are transmitted with sinusoidal trajectories having a $2\pi/g$ pitch, as shown in Fig. 7.1-1.

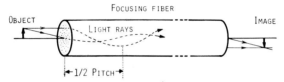

FIGURE 7.1-1.   Imaging properties of focusing fiber. Light rays are transmitted with sinusoidal trajectories having a pitch of $2\pi/g$.

Gas lenses studied as light transmission guides have been seen to show such behavior.[3,4] Subsequently, glass focusing fibers such as SELFOC fibers and rods were investigated as DI media.[5] In such cases, electromigration was employed in the formation of DI rods. Electromigration is a diffusion process in which an electric field assists migration of ions into the substrate.

Studies have also been conducted on plastic materials engineered for light transmission guides in which diffusion polymerization and other methods were employed.[6] In this chapter such DI forming methods will be reviewed, and ways to optimize the index profile will be specified.

7.2 DIFFUSION INTO OR OUT OF SUBSTRATE

One DI forming method involves the use of diffusion phenomena. If particles which contribute to increasing the index diffuse into or out of the substrate, the spatial distribution of the refractive index is formed. The fundamental mathematics of diffusion will be discussed in this section, and some examples of its utilization will be examined in later sections.

If the concentration of particles at work is expressed by $N(r, t)$ with position vector $\mathbf{r}$ and time t, the well-known diffusion equation[7] given below comes into play:

$$\partial N(\mathbf{r}, t)/\partial t = \nabla \cdot [D\nabla N(\mathbf{r}, t)], \tag{7.2-1}$$

where D is a diffusion coefficient that is not necessarily constant with respect to r and t but does depend on $N(r, t)$ itself. If D is a constant, the diffusion equation reduces to

$$\partial N(\mathbf{r}\ t)/\partial t = D\nabla^2 N(\mathbf{r}, \tau). \tag{7.2-2}$$

When the diffusion coefficient is constant,

$$\partial N(z, t)/\partial t = D[\partial^2 N(z, t)/\partial z^2] \tag{7.2-3}$$

for diffusion from planar source in one dimension. The solution of Eq. (7.2-3) is easily obtained as

$$N(z, t) = \frac{N_0}{2\sqrt{\pi Dt}} \exp(-z^2/4Dt), \tag{7.2-4}$$

where N_0 is the value at $z = 0$ and $t = 0$. Expression (7.2-4) vanishes everywhere except at $z = 0$ for $t = 0$, and then tends to $N_0\ \delta(x)$.

Next, considering a circular cylinder where diffusion is radially symmetric, concentration can be seen to be a function of radius r and time t only. The solution of the diffusion equation is given by

$$\frac{\partial N(r, t)}{\partial t} = \frac{1}{r}\frac{\partial}{\partial r}\left(rD\frac{\partial N(r, t)}{\partial r}\right), \tag{7.2-5}$$

Again, assuming constant D, and working toward a solution of Eq. (7.2-5) by using the method for separating variables,

$$N(r, t) = R(r)\exp(-Dt\alpha^2). \tag{7.2-6}$$

This is the case provided $R(r)$ is function of r only. Thus

$$\frac{d^2R}{dr^2} + \frac{1}{r}\frac{dR}{dr} + \alpha^2 R = 0 \tag{7.2-7}$$

is satisfied. This is Bessel's equation of order zero, and its solution is given by

$$R(r) = CJ_0(\alpha r). \tag{7.2-8}$$

If we assume that the initial concentration is $f(r)$ and zero at surface $r = r_0$ every time,

$$N(r_0, t) = 0 \qquad \text{for} \quad t > 0 \qquad\qquad (7.2\text{-}9)$$

$$N(r, 0) = f(r) \qquad \text{for} \quad t = 0. \qquad\qquad (7.2\text{-}10)$$

The solution is given by

$$N(r, t) = \sum_{m=1}^{\infty} C_m J_0(\alpha_m r) \exp(-Dt\alpha_m^2), \qquad\qquad (7.2\text{-}11)$$

provided α_m is the mth root of

$$J_0(\alpha_m r_0) = 0. \qquad\qquad (7.2\text{-}12)$$

Here, the initial concentration $f(r)$ has been expanded in a series of Bessel functions of order zero:

$$f(r) = \sum_{m=1}^{\infty} C_m J_0(\alpha_m r). \qquad\qquad (7.2\text{-}13)$$

Proceeding, the C_m's can be arrived at by integrating Eq. (7.2-13) after multiplying $r J_0(\alpha_m r)$ and using the orthogonality of Bessel functions in the manner

$$C_m = \frac{\displaystyle\int_0^{r_0} r J_0(\alpha_m r) f(r) \, dr}{\displaystyle\int_0^{r_0} r \{J_0(\alpha_m r)\}^2 \, dr} \qquad\qquad (7.2\text{-}14a)$$

Finally, since the denominator is given by $\frac{1}{2} r_0^2 J_1^2(\alpha_m r_0)$, the solution is

$$N(r, \tau) = \frac{2}{r_0^2} \sum_{m=1}^{\infty} \frac{J_0(\alpha_m r)}{J_1^2(\alpha_m r_0)} \exp(-Dt\alpha_m^2) \int_0^{r_0} f(r) J_0(\alpha_m r) r \, dr. \quad (7.2\text{-}14b)$$

This situation corresponds to diffusion of matter from the cylinder into the outer medium where the diffusion constant is infinitely large.

On the other hand, if the initial condition is such that

$$N(r_0, t) = N_0 \qquad \text{for} \quad t > 0 \qquad\qquad (7.2\text{-}15)$$

$$N(r, 0) = f(r) \qquad \text{for} \quad t = 0, \qquad\qquad (7.2\text{-}16)$$

the process is considered to involve an inward diffusion of particles from an outer medium having a constant concentration N_0 at $r = r_0$. The solution is

$$N(r, t) = N_0 \left[1 - \frac{2}{r_0^2} \sum_{m=1}^{\infty} \frac{1}{\alpha_m} \frac{J_0(\alpha_m r)}{J_1^2(\alpha_m r_0)} \exp(-Dt\alpha_m^2) \right]$$

$$+ \frac{2}{r_0^2} \sum_{m=1}^{\infty} \frac{J_0(\alpha_m r)}{J_1^2(\alpha_m r_0)} \exp(-Dt\alpha_m^2) \int_0^{r_0} f(r) J_0(\alpha_m r) r \, dr. \quad (7.2\text{-}17)$$

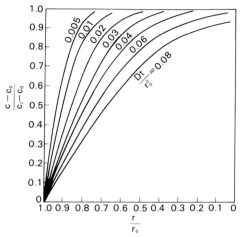

FIGURE 7.2-1. Distribution of out-diffusion in cylindrical rod. (After Kitano *et al.*[8])

An example of the numerical solution of Eq. (7.2-17), provided $f(r) = N_1$(const.) throughout the cylinder for $t = 0$, has been provided by Kitano *et al.*[8] to describe the ion exchange process for SELFOC rods (see Fig. 7.2-1).

At this point, a comment on the relation between the refractive index and concentration of matter as treated above is in order. The refractive index $n(r, t)$ can be related to the particle concentration by Clausius–Mossoti's relation

$$n^2 = n_2^2(1 + \tfrac{2}{3}\alpha N)/(1 - \tfrac{1}{3}\alpha N),\qquad(7.2\text{-}18)$$

where n_2 is the substrate index and α the polarizability of diffused matter. It then becomes possible to determine the contribution of diffused matter to the

TABLE 7.2-I

REFRACTIVE INDICES OF SILICATE GLASSES
DOPED WITH A MONOVALENT-ION OXIDE.[8]

	Glass Composition	
Modifying oxide (R_2O)	SiO_2 70 mol% R_2O 30 mol%	SiO_2 60 mol% CaO 20 mol% R_2O 20 mol%
Li_2O	1.53	1.57
Na_2O	1.50	1.55
K_2O	1.51	1.55
Rb_2O	1.50	1.54
Cs_2O	1.50	1.54
Tl_2O	1.83	1.80

TABLE 7.2-II

REFRACTIVE INDICES OF SILICATE
GLASSES DOPED WITH A
DIVALENT-ION OXIDE.[8]

Modifying oxide (RO)	Glass Composition
	SiO$_2$ 60 mol% RO 40 mol%
PbO	1.81
BaO	1.68
CdO	1.64
SrO	1.61
CaO	1.59
ZnO	1.58
BeO	1.54
MgO	1.51

index distribution after the distribution of particles has been estimated through the use of Eq. (7.2-18).

Data on the contribution of metal ions to refractive indices have been reported.[8] Reference to these data was made in the compilation of Tables 7.2-I and 7.2-II, which give the refractive indices of silicate glasses that have been treated with various modifying oxides.

7.3. ELECTROMIGRATION

Electromigration is a technique by which ions migrate with the help of an electric field. Since diffusion takes place at the same time, this process is sometimes called field-assisted diffusion. Electromigration is often utilized in the fabrication of a dielectric waveguide by migrating ions into a planar substrate in such a way that an increase in the index is contributed to.[9] It is also effective for assisting migrating ions deeply into the substrate to enable fabrication of a planar microlens (this will be discussed in a later section). We shall now describe the fundamental concept underlying electromigration.

The concentration of migrated ions obeys the diffusion equation, which includes a drift term, given as

$$\partial N(\mathbf{r}, t)/\partial t = D\nabla^2 N(\mathbf{r}, t) + \mu\nabla\varphi \cdot \nabla N(\mathbf{r}, t)$$
$$+ \mu N(\mathbf{r}, t)\nabla^2\varphi, \qquad (7.3\text{-}1)$$

where φ is the applied electric potential and μ mobility. If the drifting ions do not contribute significantly to perturbation of the electric field, the applied field maintains its initial distribution and the problem remains much simpler.

The simplest case, in which the electric field is uniform throughout the substrate and electromigration takes place in one dimension, will now be considered, that is

$$\nabla \varphi = -E \quad \text{(const)}. \qquad (7.3\text{-}2)$$

In this case, the equation reduces to

$$\varphi(z) = \varphi_0 - Ez. \qquad (7.3\text{-}3)$$

If the drift term [second term in Eq. (7.3-3)] dominates the diffusion term, the approximation

$$\partial N/\partial t = D\,\partial^2 N/\partial z^2 - \mu E\,\partial N/\partial z \qquad (7.3\text{-}4)$$

can be made

$$\partial N/\partial t = -\mu E\,\partial N/\partial z. \qquad (7.3\text{-}5)$$

The solution of this equation is given by

$$N \begin{cases} = N_0 & \text{for} \quad x < \mu E t \\ = 0 & \text{for} \quad x > \mu E t. \end{cases} \qquad (7.3\text{-}6)$$

In general cases the electric field is more complex and diffusion takes place simultaneously, as will be seen in the DI planar microlens process described in a later section.

7.4. DISTRIBUTED-INDEX PLASTIC ROD LENSES

A new method was developed by which plastic focusing fibers with improved features were fabricated. In this section, the process of measuring and controlling the index profile to minimize imaging aberrations will be described for those fibers. Imaging characteristics will also be described, and applications for the fibers will be discussed.

DI fibers or rods made of plastic materials are considered to have several merits, including (1) flexibility and nonbreakability; (2) a large refractive-index difference of $\Delta = 5\%$; (3) reasonable loss (a few decibels per meter); (4) light weight; (5) the possibility of low chromatic aberration; and (6) economy. Fabrication of plastic focusing fibers was first reported by Ohtsuka[10] who used the diffusion polymerization technique. Extensive studies of plastic focusing fiber have been carried out by Ohtsuka, as well as the present authors and associates.

The present authors have investigated the optimum plastic fiber fabrication conditions for minimization of image aberrations.[11] General image transmission characteristics as well as actual applications for these fibers have also been studied.[12]

A. FABRICATION

DI plastic rods with a nearly parabolic index distribution were obtained by a monomer exchange diffusion process occurring when a polymerized soft plastic rod with a higher refractive index was placed in a bath having a lower refractive-index monomer (see Fig. 7.4-1).

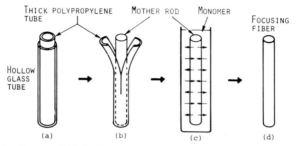

FIGURE 7.4-1.   Process of fabrication for plastic focusing fibers: (a) prepolarization; (b) cutting out; (c) diffusion; (d) polymerization and polishing.

In the first step, the plastic monomer with a higher refractive index was put in a cylindrical polypropylene tube and polymerized for 100 to 120 min at 80°C, as is shown in Fig. 7.4-1a. A thinner polypropylene tube had previously been used to mold the monomer during polymerization, but it was barely possible to obtain rods without imperfections. A thicker polypropylene tube was reshaped to guarantee a more perfect roundness by using the equipment shown in Fig. 7.4-2. It was then heated to about 100°C and pressurized in a nitrogen ambient medium at 3.5 atm so that it would fit the cylindrical glass tube.

In the second step, a soft mother rod was isolated by cutting off the polypropylene tube, as is shown in Fig. 7.4-1b. In this way, 1–3-mm diameter rods, 150 mm long without any flaws or cracks were obtained.

In the third step, the soft rod was placed in a bath at $80 \pm 0.2°C$ of a plastic monomer having a lower refractive index. This was done for a time T_2 varied between 5 and 25 min. This step is illustrated in Fig. 7.4-1c. An exchange

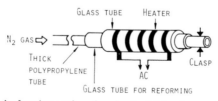

FIGURE 7.4-2.   Sketch of equipment for reforming the thick polypropylene tube used as a mold in the fabrication of a soft mother rod having a higher refractive index.

diffusion process then took place, and a DI rod with a nearly parabolic index distribution was obtained.

In the last and fourth step, the rod was completely polymerized for several hours at 80°C. Finally, the rod ends were polished successively by using polishing powders having particle sizes, in order, of 16, 7.9, 5, and 0.3 mm. A fiber scope was produced as a result.

Two pairs of plastic materials were used in our experiments, as can be seen in Table 7.4-I. The first pair consisted of diallyl isophthalate (DAI) as the higher-index plastic and methylmethacrylate (MMA) as the lower one and showed a large acceptance angle up to 55°. The second pair was diethylene glycol bis allyl carbonate (CR39) and 1,1,3-trihydroperfluoropropyl methacrylate (4FMA),[13] and resulted in fibers of very small chromatic aberration.

Rod samples 1–3 mm in diameter and 150 mm long were fabricated by the process just described. Such dimensions are satisfactory for a number of applications, for example, medical equipment. Fiber loss is no more than a few decibels per meter and does not constitute a problem for scopes of the size we are concerned with.

TABLE 7.4-I

PLASTIC MATERIALS FOR FOCUSING FIBERS[a]

Materials	Case I	Case II
Higher-index plastic P_1	DAI (1.57)	CR39 (1.51)
Lower-index plastic P_2	MMA (1.49)	4FMA (1.42)

[a] Numbers in parentheses are refractive indices of polymers.

B. CONTROL OF REFRACTIVE-INDEX DISTRIBUTION

Fiber scopes utilized in medical applications (e.g., for early detection and observation of congenital defects in fetuses)[14] must have a resolving power of 10 mm^{-1}. It was estimated by means of the following theoretical discussion that for a fiber 150 mm long to achieve such a resolution, the fourth-order coefficient h_4 of the refractive-index distribution in Eq. (7.4-1) must be kept below 3.

A quantitative measure of the image transmission characteristics for an arbitary fiber can be defined as the spatial frequency u at which the amplitude of the optical transfer function obtained from the spot diagram of the fiber[15] equals half its maximum. This can be expressed as

$$\Delta u = 153 g n^3(0)[(1 + gb)|h_4 - \tfrac{2}{3}|]^{-1} \tag{7.4-1}$$

where $n(0)$ is the refractive index at the center axis and b the length of the fiber.

To increase the value of Δu, h_4 must be nearly equal to $\frac{2}{3}$. For example, if $\Delta u = 10$ mm^{-1} for a fiber 150 mm long with $g = 0.2$ mm^{-1} and $n(0) = 1.55$, $\left(h_4 - \frac{2}{3}\right)$ must be within 0.37.

When two kinds of particles (molecules and ions, for example) inside and outside a cylindrical medium are mutually diffused, the density distribution for each particle in the medium can be obtained by solving the associated diffusion equation[16] introduced in Section 7.2. It is then possible to derive the dielectric index distribution of the diffused particles. If n_1 and n_2 are the refractive indices of the first and second particles, respectively, the index coefficients g, h_4, and h_6 are then given by

$$[n_1/(n_1 - n_2)]^{1/2}gr_0 = [-2I_2(DT_2/r_0^2)]^{1/2}, \qquad (7.4-2)$$

$$\left(\frac{n_1 - n_2}{n_1}\right)^{k-1} h_{2k} = \frac{I_{2k}(DT_2/r_0^2)}{2^{k-1}[-I_2(DT_2/r_0^2)]^k} \qquad \text{for} \quad k = 2, 3, \qquad (7.4-3)$$

with

$$I_{2k}\left(\frac{DT_2}{r_0^2}\right) = \frac{(-1)^k}{2^{2k}(k!)^2} \sum_{m=1}^{\infty} \frac{2j_{0m}^{2k-1}}{J_1(j_{0m})} \exp\left[-j_{0m}^2\left(\frac{DT_2}{r_0^2}\right)\right], \qquad (7.4-4)$$

where r_0, D, T_2, J_1, and j_{0m} are, respectively, the fiber radius, diffusion constant, diffusion time, first-order Bessel function, and mth root of the equation $J_0(x) = 0$. Figure 7.4-3 shows $[n_1/(n_1 - n_2)]^{1/2}gr_0$ and $[(n_1 - n_2)/n_1]^{k-1}h_2^k$

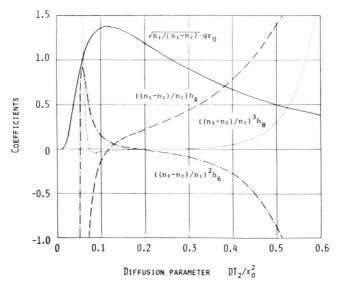

FIGURE 7.4-3. Refractive-index distribution coefficients as functions of the diffusion parameter DT_2/r_0^2.

as functions of the diffusion parameter DT_2/r_0^2. The second-order coefficient g has $g_{max} = 1.38[(n_1 - n_2)/n_1]^{1/2}/r_0$ when the fourth-order coefficient h_4 approaches zero at the same time, assuring a low level of aberration. The diffusion constant D is obtained from the maximum g_{max} and its diffusion time T_2 as

$$D = 0.221(n_1 - n_2)/n_1 T_2 g_{max}^2. \qquad (7.4\text{-}5)$$

During our experimentation with the fabricated plastic rods, the optimum diffusion time after which the focusing constant g reached its maximum (and h_4 contracted) varied within the 10- to 20-min range for the previous fabrication condition. However, due to the chemical reaction of the plastic monomer the diffusion time T_2 could not exceed 10 min. This problem was overcome by shifting the mother rod to another bath of the same lower-index monomer at regular periods of time. Temperatures for the lower-index monomer during the diffusion are plotted in Fig. 7.4-4 and show periodic behavior in relation to the diffusion time.

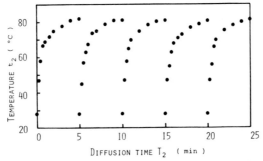

FIGURE 7.4-4.   Measured temperatures t_2, for lower-index monomer shown versus diffusion time T_2.

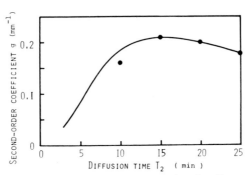

FIGURE 7.4-5.   Second-order coefficient g of a plastic focusing fiber prepared by repeated diffusion method shown versus diffusion time T_2. Experimental results are indicated by dots, and theoretical ones by the solid curve.

Since the diffusion constant D was dependent on the diffusion temperature, for general use a mean diffusion constant had to be determined. Focusing constant g (obtained through the bath-shifting diffusion process previously mentioned) was plotted versus diffusion time in Fig. 7.4-5. The value for g_{max} was estimated as 0.21 mm^{-1} when the associated diffusion time T_2 equaled nearly 15 min. Substitution of these values into Eq. (7.4-5) then led to $\bar{D} = 2.8 \times 10^{-6}$ cm^2/sec, where we have assumed that $(n_1 - n_2)/n_1 = 0.05$. Using these values, and with the help of Fig. 7.4-3, the continuous curve in Fig. 7.4-5 was drawn.

C. Measurement of Index Distribution

The refractive-index distribution of the fiber samples was measured using a differential interferometric microscope, as will be detailed in Chapter 8. A fiber rod sunken in matching oil was placed in front of a light source so that light beams crossed the sample perpendicularly to its central axis. The wavefront W_2 sheared by an arbitrary distance s was interfered differentially from wavefront W_1 and the resulting interference pattern was observed through the microscope.

A cylindrical fiber sample was first used, as shown in Fig. 7.4-6a. The interference pattern associated with a cylindrical sample having a refractive index distribution in the y direction, as in Fig. 7.4-6b, represents the optical pathlength differentiated with respect to the transverse x axis.

The extent of the nearly parabolic portion of a refractive-index distribution and abrupt changes in the refractive index of the fiber are inferred from this pattern. The profile of the mean refractive index, obtained by averaging the refractive index distribution in the y direction, is shown in Fig. 7.4-7 for soft samples. This method is most convenient because it allows nondestructive measuring. For cylindrical samples, however, correction is necessary to obtain the focusing constant g and the refractive index at the center axis of the fiber.

Next, the fiber rod was sliced along the center axis into a thin plate before embedding into matching oil for more precise measurement. This is illustrated in Fig. 7.4-8. The differentiated interference pattern shown in Fig. 7.4-9 corresponds approximately to the refractive-index distribution differentiated with respect to the transverse x axis, where the linear portion near the center has a nearly parabolic index distribution.

The refractive-index profile was obtained by integrating the pattern in Fig. 7.4.9 with respect to the transverse axis, as can be seen in Fig. 7.4-10. The refractive indices at the central axis $n(0)$ and at the edge were estimated as $n(0) = 1.545$ and $n(a) = 1.515$. The coefficient g in the parabolic term was valued at 0.207 mm^{-1}, and the fourth-order coefficient h_4 was determined as 3 near the center. The value $g = 0.207$ mm^{-1} is nearly equal to the maximum

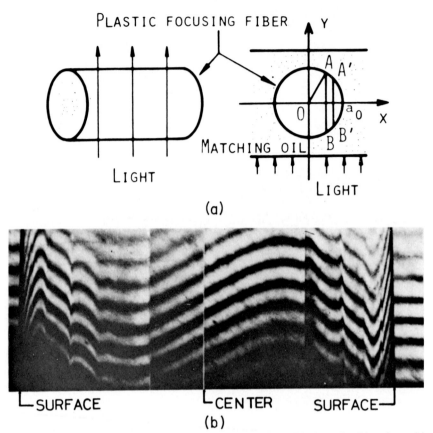

FIGURE 7.4-6.   (a) Cross section of fiber row. The light ray passing through AB interferes with that through $A'B'$. (After K. Iga and N. Yamamoto[7].) The matching oil which has the same index as the sample is used to minimize the refraction of light at the sample boundary. (b) Differentiated interference pattern associated with round plastic fiber sample.

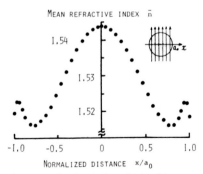

FIGURE 7.4-7.   Profile of mean refractive index obtained by averaging the refractive-index distribution in the y direction. The transverse axis x/a_0 represents the distance from the center axis of the normalized fiber sample.

118

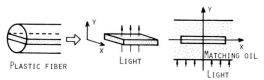

FIGURE 7.4-8.   Thin plate prepared to achieve more precise measurement of refractive-index distribution for plastic focusing fibers.

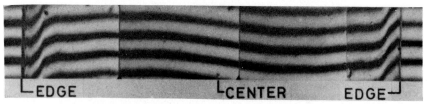

FIGURE 7.4-9.   Differentiated interference pattern associated with the thin plate. Prepared by slicing a round fiber sample along the center axis.

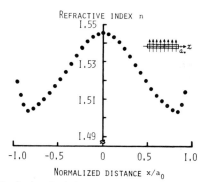

FIGURE 7.4-10.   Profile of refractive index measured using a thin plate. The transverse axis x/a_0 represents the normalized distance from the central axis of the fiber sample.

estimated theoretically by the use of Eq. (7.4-2), with $(n_1 - n_2)/n_1 = 0.05$ and $r_0 = 1.5$ mm. The spatial frequency for a fiber 25 mm long was then estimated using Eq. (7.4-1), resulting in $\Delta u = 8$ mm^{-1}). Here, $g = 0.207$ mm^{-1}, $h_4 = 3$, and $n(0) = 1.545$ were used. The estimated value, $h_4 \sim 3$, was almost satisfactory for image transmission associated with this fiber rod length.

D. IMAGING CHARACTERISTICS

The resolving power of the fiber samples was measured by observing Ronchi grating images, as shown in Fig. 7.4-11, which shows a 12 mm^{-1} grating image (with magnification 50% and white source light) of a fiber rod

FIGURE 7.4-11.   Image of Ronchi grating ($12\ mm^{-1}$) observed using plastic focusing fiber 25 mm long.

25 mm long produced by a DAI–MMA diffusion. This fiber sample was different from the one used to measure the index distribution in the previous section and had better imaging properties.

The estimated resolving power of this fiber was $45\ mm^{-1}$ which means that a fiber 150 mm long can be obtained with a resolving power of 5 to $6\ mm^{-1}$ and that it will be capable of further improvement through the use of monochromatic light illumination. The resolving power is limited by chromatic and fourth-order aberrations. Chromatic aberration decreased for fibers obtained from a CR39–4FMA pair.

Next, attention was turned toward some simple examples of plastic focusing fibers usable for imaging applications. Figure 7.4-12 shows a scope composed of a short plastic fiber rod with a large NA and a long rod with small NA Figure 7.4-12 shows a plastic fiber rod where one of the ends is rounded to decrease chromatic aberration. Figure 7.4-12 illustrates the serial composition of sevveral plastic fiber rods for magnifying images.

Some simple imaging systems utilizing plastic fiber rods come easily to mind. For example, Fig. 7.4-13 shows a simple fiber scope with an eyepiece. Figure 7.4-13 shows a system consisting of a plastic fiber scope connected to a TV monitor camera, which is suitable for many applications. In medical applications, for instance, it may be used in a device tailored to the early detection and observation of congenital defects in fetuses, tracheal diseases, etc. Finally, Fig. 7.4-13 shows a plastic fiber lens connected to a fiber bundle.

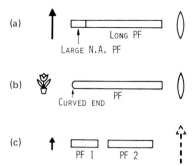

FIGURE 7.4-12. Arrangement of plastic fibers (PF) for imaging. See text for discussion.

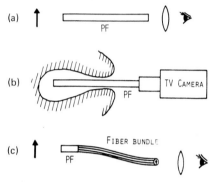

FIGURE 7.4-13. Imaging systems using plastic focusing fiber. See text for discussion.

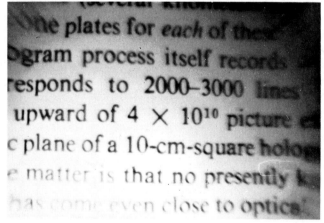

FIGURE 7.4-14. Image of letters $170 \times 170 \, \mu$m in size observed using 25-mm-long fiber obtained from DAI–MMA diffusion.

FIGURE 7.4-15.   TV monitoring system using plastic focusing fiber.

TABLE 7.4-II

PRESENT STATUS OF PLASTIC FOCUSING FIBER

	Length (mm)	Diameter (mm)	g (mm^{-1})	h_4	Resolution/mm
One target	150	1–3	0.2	1.5–2	10
Present status	150	1–3	0.2	5	5

As a real application, letters 170 × 170 mm in size were transmitted through a plastic fiber rod 25 mm long fabricated from a DAI–MMA pair. The transmitted image is reproduced in Fig. 7.4-14. Furthermore, a prototype system composed of a plastic fiber scope connected to a TV monitor was constructed in the manner shown in Fig. 7.4-15. This type of a system is likely to have wide applicability in different fields.

As a final comment, the present status of research on plastic focusing fibers as reported here is presented in comparative form in Table 7.4-II along with a target status as previously fixed by the authors. The resolving power of 5 mm^{-1} that is attainable at present for 150 mm needs to be further improved at an early date.

E. CONCLUDING REMARKS

A new method for the fabrication of plastic focusing fibers was developed, and effective control of the refractive index distribution for better image transmission was realized. Refractive-index profiles were measured through transverse differential interferometry, and the index distribution coefficients, g and h_4, were estimated. Satisfactory image transmission was seen to be possible. Among the many applications the authors foresee for plastic focusing fibers, emphasis was placed in this research upon the possibility of medical devices, such as tracheascopes and congenital defect detecting systems.

7.5 DISTRIBUTED-INDEX ROD LENS WITH OPTIMIZED DIFFUSION

A DI fiber with nearly parabolic index distribution has transmission properties favorable for optical communication[17] and image transmission.[18] Focusing fibers have been made of glass[19] or plastic materials[20,21] through diffusion exchange of ions or molecules. Changes in index distribution have been investigated,[19,22,23] and almost optimum diffusion conditions for low aberration obtained for plastic focusing fibers.[21] However, it was made clear that the fourth- and sixth-order coefficients, h_4 and h_6, in Eq. (7.2-1) cannot be simultaneously at their optimum values when only diffusion exchange is used.[22] It was confirmed empirically and experimentally that aberrations in focusing fibers which were fabricated by diffusion exchange decreased with heat treatment.[24,25] However, no theoretical consideration of both the diffusion-exchange process and successive heat-treatment process has been reported. In this section it will be made clear that h_4 and h_6 can be at their optimum values simultaneously with application of the thermal diffusion in a heat-treatment process following the diffusion-exchange process. Here we define "optimum" as the condition where the meridional ray distortion becomes minimum, i.e., $h_4 = \frac{2}{3}$ and $h_6 = -\frac{17}{45}$. This condition is sufficient for such applications as focusing lenses or view scopes.

In fabricating a focusing fiber by the diffusion method, particles (ions or monomers) which give the index variation in a cylindrical medium diffuse outside during the diffusion-exchange process. The density distribution of particles in the cylindrical medium can be obtained by solving the associated diffusion equation[22]

$$\frac{\partial N(r, t)}{\partial t} = \frac{1}{D}\left(\frac{\partial^2 N}{\partial r^2} + \frac{1}{r}\frac{\partial N}{\partial r}\right), \tag{7.5-1}$$

If we put $\rho = r/r_0$, where r_0 is the radius of the medium, which is not necessarily equal to the final radius of the fiber, then the density $N_d(\rho, T_2)$ of particles which diffuse away from the center of the cylindrical medium can be obtained as follows with the help of the assumption that the density outside the medium is zero.

$$N_d(\rho, T_2)/N_{di} = \sum_{k=0}^{\infty} I_{2k}(T_2)\rho^{2k} \tag{7.5-2}$$

with

$$I_{2k}(T_2) \equiv \frac{(-1)^k}{2^{2k}(k!)^2} \sum_{m=1}^{\infty} \frac{2j_{0m}^{2k-1}}{J_1(j_{0m})} \exp\left\{-j_{0m}^2 \frac{D_2 T_2}{r_0^2}\right\}, \tag{7.5-3}$$

where N_{di}, D_2, T_2, J_1, and j_{0m} are, respectively, the initial density $N_d(\rho, 0)$, the diffusion constant in the diffusion-exchange process, the diffusion time, the first-order Bessel function, and the mth zero of the zeroth order Bessel function J_0.[22] Figure 7.2-1 showed the change of density of diffused particle as a function of in-diffusion time T_2.

Next, we need to solve is the diffusion equation in the case where the heat treatment is applied successively after the diffusion exchange. Since particles in a cyclindrical rod do not diffuse beyond the periphery of the rod in the heat-treatment process, we can assume that the slope of density for the particles is zero at the periphery of the rod. After the heat treatment is applied for the time period T_3, the density distribution of particles that diffuse away from the center in the heat-treatment process $N_h(r, T_3)$ can be obtained by solving the diffusion equation with the help of this assumption. The density $N_h(r, T_3)$ is

$$N_h(\tau, T_3) = 2\left[\int_0^1 \rho f(\rho)\, d\rho + \sum_{n=1}^{\infty} \frac{J_0(j_{1n}\tau/\tau_0)}{J_0^2(j_{1n})} \exp\{-j_{1n}^2(D_3 T_3/\tau_0^2)\}\right.$$
$$\left. \times \int_0^1 \rho f(\rho) J_0(j_{1n}\rho)\, d\rho\right], \tag{7.5-4}$$

where $f(\rho)$, D_3, and j_{1n} are the initial density distribution in the heat-treatment process, the diffusion constant in the heat-treatment process, and the nth root of the equation $J_1 = 0$, respectively.

When heat treatment is applied after diffusion exchange, the density distribution is obtained by substituting Eq. (7.5-2) for $f(\rho)$ in Eq. (7.5-4). The refractive-index distribution is derived from the Clausius–Mosotti relation with the help of the polarization index and the density distribution of particles.[22] The index coefficients, g, h_4, and h_6, are then given by

$$\sqrt{n_1/(n_1 - n_2)}g\tau_0 = \sqrt{-2H_2(T_3)}, \tag{7.5-5}$$

$$[(n_1 - n_2)/n_1]^{k-1}h_{2k} = \{H_{2k}(T_3)/2^{k-1}[-H_2(T_3)]^k\} \quad (k = 2, 3), \tag{7.5-6}$$

with

$$H_{2k}(T_3) \equiv \frac{(-1)^k}{2^{2k}(k!)^2}\left[\sum_{n=1}^{\infty} \frac{2j_{1n}^{2k}}{J_0^2(j_{1n})} \exp\left(-j_{0n}^2 \frac{D_3 T_3}{\tau_0^2}\right) \times \int_0^1 f(\rho) J_0(j_{1n}\rho)\, d\rho\right], \tag{7.5-7}$$

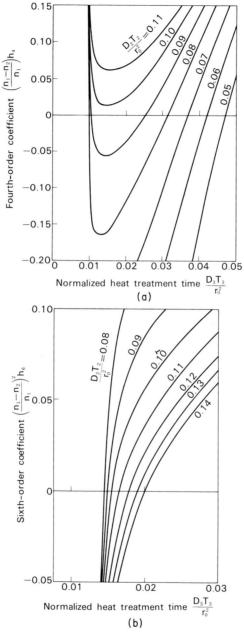

FIGURE 7.5-1. (a) Fourth- and (b) sixth-order coefficients, h_4 and h_6, in the heat-treatment process shown versus normalized heat-treatment time $D_2 T_3 / r_0^2$. The parameter $D_2 T_2 / r_0^2$ is normalized diffusion-exchange time.

where n_1 denotes the initial refractive index of the cylindrical medium in which particles exist homogeneously and n_2 that index associated with a medium without particles. The fourth and sixth-order coefficients (h_4 and h_6) after the heat-treatment process are shown in Fig. 7.5-1 as functions of the normalized heat treatment time $D_3 T_3/r_0^2$.

When meridional rays which are transmitted through focusing fibers are considered, it is seen that the fourth- and sixth-order coefficients, h_4 and h_6, must be $\frac{2}{3}$ and $-\frac{17}{45}$, respectively, for obtaining focusing fibers without aberration. Optimum diffusion-exchange and heat-treatment conditions for $h_4 = \frac{2}{3}$ and for $h_6 = -\frac{17}{45}$ can be obtained from Fig. 7.5-1a and Fig. 7.5-1b, respectively. These optimizations are shown in Fig. 7.5-2 by dashed–dotted curves, where we have assumed that $(n_1 - n_2)/n_1 = 0.01$. Furthermore, the combinations of $D_2 T_2/r_0^2$ and $D_3 T_3/r_0^2$ which satisfy $h_4 - \frac{2}{3} = 1$ and 3 are obtained from Fig. 7.5-1a. They are shown in Fig. 7.5-2 by solid curves and dashed curves, respectively. The combinations of $D_2 T_2/r_0^2$ and $D_3 T_3/r_0^2$ which satisfy $h_6 + \frac{17}{45} = 100$ and 300 which were also obtained from Fig. 7.5-1b are shown in Fig. 7.5-2 by solid curves and dashed curves, respectively. The superimposed region indicated by the arrow in Fig. 7.5-2 represents the diffusion-exchange and heat-treatment conditions for obtaining low-aberration focusing fibers where both h_4 and h_6 are small.

When only the diffusion-exchange process is applied, the optimum diffusion condition for $h_4 = \frac{2}{3}$ is $D_2 T_2/r_0^2 = 0.12$. When heat treatment is applied after diffusion exchange, the optimum diffusion-exchange and heat-treatment

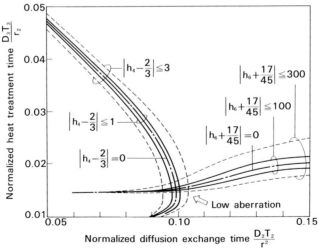

FIGURE 7.5-2. Optimum diffusion-exchange and heat-treatment conditions. Superimposed region indicated by the arrow shows optimum low-aberration values for fourth- and sixth-order coefficients, h_4 and h_6.

conditions for $h_4 = \frac{2}{3}$ and $h_6 = -\frac{17}{45}$ were found to be $D_2 T_2/r_0^2 = 0.10$ and $D_3 T_3/r_0^2 = 0.015$. As can be seen, the optimum diffusion-exchange and heat-treatment condition where aberrations caused by h_4 and h_6 were simultaneously weakened were obtained by applying the heat-treatment process in addition to the diffusion-exchange process.

This idea was introduced into the process control of DI glass rod lenses.[26] Low-aberration samples resulted. In Fig. 7.5-3 the aberration characteristic are shown as a function of diffusion time.

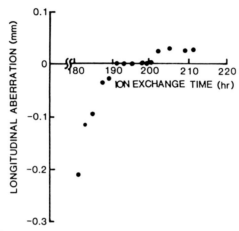

FIGURE 7.5-3. Measured longitudinal aberration of experimental selfoc lenses shown in relation to ion-exchange time, where bath temperature is 540°C. (After Miyazawa et al.[26]) The exchange time of 190 to 200 h gives us a low aberration. This indicates that the index profile is optimized at this time period.

7.6 DISTRIBUTED-INDEX PLANAR MICROLENSES

Conventional distributed (or gradient) index rod lenses[27] and microspherical lenses[28] have been widely used as optical components in optical fiber communication systems and optoelectronic systems. The planar microlens to be presented here has the same area of application as conventional lenses and has the advantage that it is monolithically produced by means of the planar technology commonly used in electronics. Although the surface is planar, a lens effect results from a dopant selectively diffused through a mask onto the planar substrate, as shown in Fig. 7.6-1. A number of lenses, or a two-dimensional array of them, can be obtained monolithically. Surface treatments such as the preparation of filters, apertures, and dielectric coatings, may be easily realized by a single batch process. Moreover, the need for adjustment of the optical axis may be alleviated. This section will present the design,

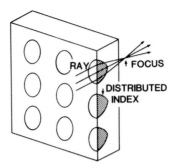

FIGURE 7.6-1.   Structure of the distributed index planar microlens array.

fabrication, and measurement of such lenses and shall comment further on the development of the planar microlens.

There are two selective diffusion methods for realizing a positive–negative planar microlens, and they depend on the characteristics of the dopant. That is, when a dopant results in an increment of the refractive index, a negative–positive mask is used. Otherwise, when a decrease in the index results from diffusion of the dopant, a positive–negative mask is applied. Prior to initial planar microlens fabrication, selective diffusion was checked to see whether it would result in an index distribution that would bring about lens effect.

First, we examined how the dopant diffused from a very small window of the mask, and how the light ray was refracted in the resultant distributed index near the window. If the window was small enough compared with the diffusion length, the window was considered to be a small hemispherical diffusion source with radius r_m. The dopant concentration at time t is given by the equation[29]

$$u(r; t) = u_0 \frac{r_m}{r} \operatorname{erfc}\left(\frac{r - r_m}{2\sqrt{Dt}}\right), \qquad (7.6\text{-}1)$$

where r is the distance from the center of the diffusion source and D the diffusion constant when D is independent of concentration.

The curvature of light ray K can also be determined as

$$|K| = |\operatorname{grad} \log n(r)|, \qquad (7.6\text{-}2)$$

where n is the refractive index of medium (see Fig. 7.6-2).[30]

Since the index increment is proportional to the dopant concentration according to Clausius–Mossotti's relation and since the derivative of Eq. (7.6-1) monotonically decreases, curvature K also decreases with respect to r. This brings about the large aberration shown in Fig. 7.6-2. Such a diffusion process, therefore, cannot be applied to the fabrication of planar

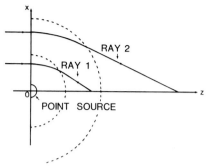

FIGURE 7.6-2.   Light-ray trajectory in a distributed-index medium achieved by diffusion from a small hemispherical source.

microlenses as long as the diffusion process is linear. However, it may be possible to create a low-aberration lens when the diffusion constant has concentration dependence.

Next, a case was examined where the radius of the window r_m was not so small compared with the diffusion length. Assuming a boundary condition where the dopant flowed constantly into a substrate through the window, the dopant concentration was obtained by integrating Green's function of diffusion[31] such that

$$u(\tilde{r}, \tilde{z}; \tau) = J_0 \int_0^1 \int_0^{2\pi} \tilde{r}' \, d\tilde{r}' \, d\theta' \, G(\tilde{r}, \theta, \tilde{z}, \tilde{r}', 0, \tilde{z}'; \tau), \qquad (7.6\text{-}3)$$

$$G(\tilde{r}, \theta, \tilde{z}, \tilde{r}', \theta', \tilde{z}'; \tau) = r_m \operatorname{erfc}(\tilde{R}/\sqrt{\tau})/(2\pi^{3/2} D\tilde{R}), \qquad (7.6\text{-}4)$$

$$\tilde{R} = (\tilde{r}^2 - 2\tilde{r}\tilde{r}' \cos \theta' + \tilde{r}'^2 + \tilde{z}^2). \qquad (7.6\text{-}5)$$

Here we have used a cylindrical coordinate system, where z is the distance from the surface in the depth direction, r in the radial direction, and θ in the azimuthal direction though in the left-hand side of Eq. (7.6-3) this is omitted by the assumption of axial symmetry; the overtilde means normalization by r_m; and t is normalized diffusion time, i.e., $t = Dt/r_m^2$; and J_0 denotes the flow at $z = 0$.

Figure 7.6-3 shows light-ray trajectories in resultant distributed indices. These are associated with a typical normalized diffusion times. It was also assumed that the index increment was proportional to the dopant concentration. The maximum was chosen for the normalized index difference $\Delta n/n = 0.1$. To obtain the ray trajectory we used the ray equation[32]

$$\ddot{r} + \frac{1}{n^2}(1 + \dot{r}^2)\dot{r}\frac{\partial}{\partial z}\left(\frac{1}{2}n^2\right) - \frac{1}{n^2}(1 + \dot{r}^2)\frac{\partial}{\partial r}\left(\frac{1}{2}n^2\right) = 0, \qquad (7.6\text{-}6)$$

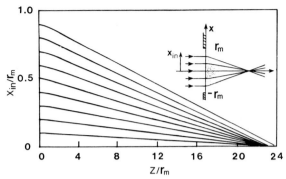

FIGURE 7.6-3. Estimated light-ray trajectories for various normalized diffusion time, where $\Delta n/n = 0.1$ and $Dt/r_m^2 = 0.56$. Here x_{in} and z/r_m indicated the incident ray position and the distance

where the overdot means the derivative with respect to z. Equations (7.6-3) and (7.6-6) were numerically integrated.

Looking at Fig. 7.6-3, it can be seen that the aberration becomes small when the normalized diffusion time Dt/r_m^2 is larger than 0.36. The preferable diffusion time for creating the planar microlens is $Dt/r_m^2 \gtrsim 0.4$. Furthermore the focal length in the substrate is estimated to be about 20 times the mask radius r_m and the numerical aperture NA $= \frac{1}{20} \times n_2$, where n_2 is the index of substrate and the NA is considered to be proportional to $\Delta n/n$.

When a lower index dopant was diffused from around the disk, the density of the dopant was expressed by the difference between the density for the uniform diffusion and that for the diffusion through the window, i.e.,

$$u(\tilde{r}, \tilde{z}; \tau) = J_0 \left[\int_0^{-} \int_0^{2\pi} \tilde{r}' \, d\tilde{r}' \, d\theta' \, G(\tilde{r}, \theta, \tilde{z}, \tilde{r}', \theta', \tilde{z}'; \xi) \right.$$
$$\left. - \int_0^1 \int_0^{2\pi} \tilde{r}' \, d\tilde{r}' \, d\theta' \, G(\tilde{r}, \theta, \tilde{z}, \tilde{r}', \theta', \tilde{z}'; \tau) \right]. \qquad (7.6-7)$$

Since the index distribution in the radial direction is determined by the second term of Eq. (7.6-7), which is the same as Eq. (7.6-3), and because the radial index distribution contributes mostly to the ray trajectory, the difffusion conditions mentioned above are also applicable in the latter case.

A planar microlens was then fabricated with two different refractive index plastics by means of monomer exchange diffusion[33-35]. Plastics are suitable for making larger-diameter lenses, since they have large diffusion constants at relatively lower temperatures, i.e., 100°C. The fabrication time necessary for making a planar microlens with a radius of 0.5 mm is given in Table 7.6-I for various materials.

TABLE 7.6-I

SUMMARY OF FABRICATION TIMES

Materials	W_n/n	D (m^2/sec)	t (sec)^a
Plastics (DAI–MMA)	0.05	3×10^{-10}	3×10^2
Glass (Tl) ion-exchange	0.05	4×10^{-13}	9×10^4
Glass (Tl) electromigration	0.05	—	$3 \times 10^{4\,b}$

^{a} $t = (r_m^2/D) \times 0.4$
^{b} Experimental data with radius of 0.6 mm.

In our initial experiment we used DAI (diallyl isophtalate, $n = 1.570$) as a higher-index substrate and MMA (methyl methacrylate, $n = 1.494$) as a lower-index dopant which was to be diffused from around the disk mask. First, we put DAI monomer containing 4% BPO (benzoil peroxide) as an initiator into a cylindrical polyethylene case with a diameter of 24 mm and depth of 7 mm. After placing a cover on the case we heated it at a constant 80°C for 85 min to achieve partial polymerization. A planar substrate with a thickness of 3 to 4 mm was obtained.

Next, a mask (a glass disk of 3.6 mm diameter and 1 mm thickness) was put on the DAI substrate which was in contact with the polyethylene case and then placed in an MMA bath at 80°C for 55 min. The MMA diffusion took place from around the mask, and the index distribution was formed. After the diffusion process was finished, the resultant planar microlens was completely polymerized by heating at 70°C for 24 hr. Then, a plastic planar microlens 3.6 mm in diameter with a focal length of 25 mm was obtained.

Figure 7.6-4 shows the fabricated planar microlens. Magnified letters can be seen through the lens at its central part where the 3.6 mm mask was placed.

When a He–Ne laser beam ($\lambda = 632.8$ mm) was illuminated on the plastic planar microlens, an Airy-like focus spot was observed with a diameter of 27 mm.

Use of a glass substrate for a planar microlens brings about the possibility of precise mask formation through implementing photolithographic techniques. This use enables expanded application of the planar microlens, and in order to create an index distribution in the glass substrate, it is possible to use the ion-exchange technique.[36] That is, oxides of monovalent cations in a glass substrate can be changed to other monovalent cations from molten salt, thus bringing about differences in the refractive index.

The substrate used in a preliminary experiment was planar BK7 glass having 20×20 mm dimension and 5 mm thick. Titanium 2 μm thick was sputtered on it as a mask. Circular windows of $2r_m = 0.6$ mm with a pitch of 1.5 mm were opened on the metal mask by means of photolithography. The masked substrate was immersed in molten salt for 165 hr, during which time

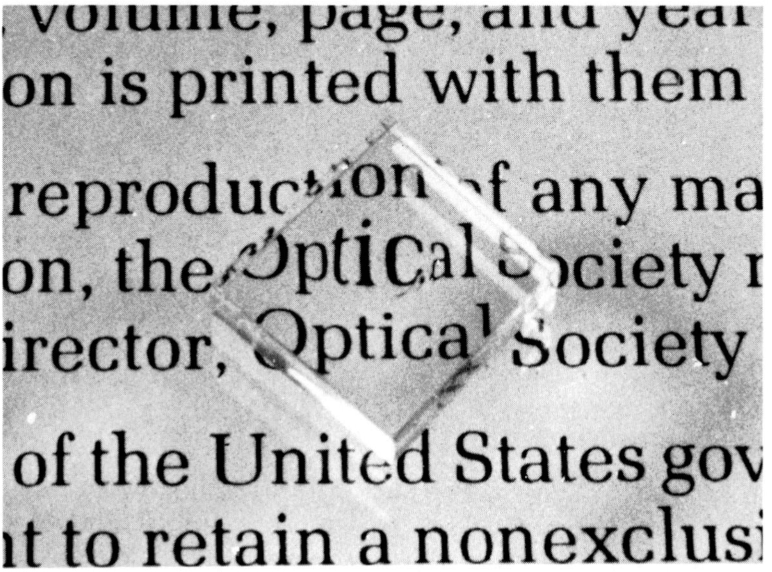

FIGURE 7.6-4.   Plastic planar microlens. Magnified letters can be seen at the center where a 3.6 mm disk mask was placed.

ion exchange took place and the distributed index was formed near the windows of the mask.

We chose Tl^+ as the dopant which could bring about a higher index and which could be exchanged for K^+ or Na^+ in a glass substrate.[36] Molten salt which included thallium sulfate was then used as a source for the dopant. Diffusion time was chosen so as to provide enough of a gradient in the refractive index. According to theoretical calculations, the normalized diffusion time needs to be in the range $0.4 < Dt/r_m^2$. We have estimated that $D = 4 \times 10^{-13} \, m^2/sec$ for Tl^+ in the glass from extrapolating the data in Refs. 36 and 37, when $r_m = 0.3$ mm and Dt/r_m^2 is larger than 2.6.

The lens effect could be observed where mask windows existed. The focal length was 9.4 mm, which was measured for the configurations shown in Fig. 7.6-5. Such a measurement does not agree with the definition of the focal length of a thick lens. However, the difference is small since the depth of the distributed index region is small ~0.4 mm) compared with the focal length. The diameter for each lens was extended to 1.2 mm, due to the diffusion in the radial direction. According to the results of theoretical analysis where it was assumed that the diffusion constant was independent of concentration, the aberration should be large near the edge of the mask window (see Fig. 7.6-3). On the contrary, though, the lens showed small aberration even outside the mask (the lens diameter was twice larger than the mask radius). This

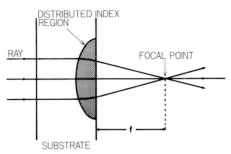

FIGURE 7.6-5.   Method for measuring focal length. Collimated light is incident on the back surface, and the distance from the front surface to the focused point is measured as focal length.

experimental result is considered to be explained by the concentration dependence of the diffusion.

Real images made from an array of planar microlenses are shown in Fig. 7.6-6. The field angle $2\theta_e$ of this sample was as wide as 1/8 rad. The spot diameter of the focused beam was observed versus the radius of an aperture which was placed just in front of the microlens. The experimental setup is shown in Fig. 7.6-7. If the lens did not have any aberration, the spot diameter was limited by the diffraction described by equation[38]

$$D_s = 2.44f\lambda/2r_a \qquad (7.6\text{-}8)$$

or

$$D_s = 1.22\lambda/NA, \qquad (7.6\text{-}9)$$

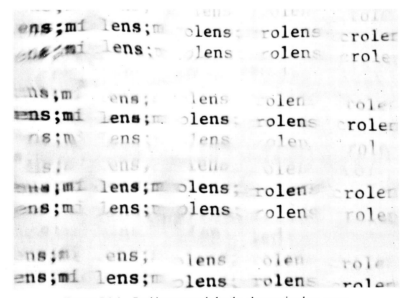

FIGURE 7.6-6.   Real images made by the planar microlens array.

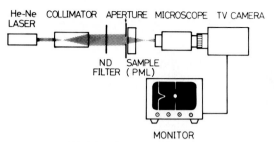

FIGURE 7.6-7 Experimental setup to measure focused spot diameter versus aperture radius. The sample is illuminated by a plane wave with an aperture stop. The focused spot is observed by a combination of microscope and TV monitor. (After M. Oikawa and K. Iga[14].)

where f is focal length, r_a the aperture radius, and λ the wavelength of the light source.

The diffraction limit and measured spot diameters are shown in Fig. 7.6-8. The spot diameter plots do not agree with the diffraction limit when the aperture radius is larger than 0.4 mm and the minimum spot diameter is 17 μm. This discrepancy is due to aberration, and the NA with no aberration can be determined from Eq. (7.6-9) for a minimum spot diameter of 17 μm in the following manner:

$$NA = 1.22\lambda/D_s = 0.05. \qquad (7.6\text{-}10)$$

The electromigration technique[39] is more effective for creating devices with short focal length and reduced fabrication time. By applying an electric field of 7 V/mm for 8 hr, it was possible to obtain a planar microlens with radius of 0.6 mm and focal length of 6.8 mm that was shorter than the sample from the

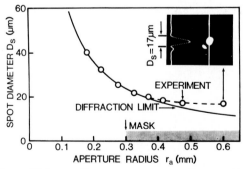

FIGURE 7.6-8. Focused light spot diameter versus aperture diameter. Plane wave from He–Ne laser was used as light source, with $f = 9.4$ mm and $\lambda = 632.8$ mm. The insert is the observed focused spot on the TV monitor shown in Fig. 7.6-7. The left-hand side trace indicates the intensity distribution of the focused spot. (After M. Oikawa and K. Iga[14].)

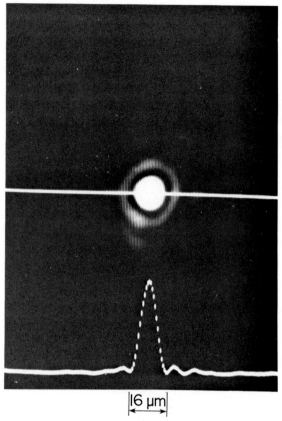

FIGURE 7.6-9. Airy-like focused spot of planar microlens fabricated by means of electromigration.

conventional ion-exchange process using a 55-mm mask. We then focused a He–Ne laser beam to 3.8 mm in diameter. The observed spot is shown in Fig. 7.6-9.

The concept behind planar microlenses was explained, and feasibilities were shown for fabrication by plastic and glass. If an NA > 0.2 is obtained with small aberration by optimizing the fabrication process, planar microlenses can be applicable to such microoptics as a multichannel coupler in single-mode fiber systems and two-dimensional arrayed lenses in optoelectronic components. The possibility of a planar microlens with a focal length as short as 2.6 mm and an NA of 0.15 has been confirmed.

TABLE 7.7-I

DI FORMING METHODS

Process	Substrate	Dopant	Form	Remarks	Reference
Ion exchange diffusion	Glass, optical crystal	Ag, Li, Tl, Cs, Pb	Rod, fiber, planar	Smooth profile	41
Electromigration	Glass, optical crystal	Ti, K, Tl	Planar	Steplike profile. fast	42
Molecular stuffing	Porous glass	Cs, Pb, Ag	Rod, planar	Large diameter and n	43
Phase separation and partial leaching	Glass having interconnected microstructures	GeO_2	Rod	Large n	44
CVD (MCVD, OVD)	Silica (SiO_2)	Ge, P, B, F, Al	Tube, rod	Multilayers	45, 46
VAD			Rod	Spiral layers	47
Plasma CVD		Si_3N_4	Planar	Multilayers	48
Diffusion polymerization		Plastic monomers	Rod, planar	Wide variety of materials, flexible	49, 50
Photocopolymerization			Rod, fiber, planar		51
Czochralski method	Semiconductor crystal (Si)	Ge	Rod	For infrared	52

7.7 OTHER METHODS

A number of methods have been reported for achieving the formation of a distributed index inside the medium, and Mukherjee has reviewed some of them.[40] Some of the methods are tabulated in Table 7.7-I along with possible materials and substrate forms.

REFERENCES

1. E. A. J. Marcatili, *Bell Syst. Tech. J.* **43**, 2887 (1964).
2. K. Iga, S. Hata, Y. Kato, and H. Fukuyo, *Jpn. J. Appl. Phys.* **13**, 79 (1974).
3. Y. Aoki and M. Suzuki, *IEEE Trans. Microwave Theory Tech.* **MTT-15**, 1 (1967).
4. Y. Suematsu, K. Iga, and S. Ito, *IEEE Trans, Microwave Theory Tech.* **MTT-14**, 657 (1966).
5. T. Uchida, M. Furukawa, I. Kitano, K. Koizumi, and H. Matsumura, *IEEE J. Quantum Electron.* **QE-6**, 606 (1970).
6. Y. Ohtsuka, *Appl. Phys. Lett.* 23, 247 (1973).
7. J. Crank, "The Mathematics of Diffusion," Clarendon Press, Oxford, 1975.
8. I. Kitano, K. Koizumi, T. Uchida, and H. Matsumura, *J. Jpn. Soc. Appl. Phys. Suppl.* **39**, 63 (1970).
9. T. Izawa and H. Nakagome, *Appl. Phys. Lett.* **21**, 584 (1972).
10. Y. Ohtsuka, *Appl. Phys. Lett.* **23**, 247 (1973).
11. K. Iga, K, Yokomori, and T. Sakayori, *Appl, Phys. Lett.* **26**, 578 (1975).
12. K. Iga, N. Yamamoto, and Y. Matsuura, in "CLEOS, 25–27 May 1976, Digest of Technical Papers," paper WD9. Optical Society of America, Washington, D.C., 1976.
13. Y. Ohtsuka, T. Senga, and H. Yasuda, *Appl. Phys. Lett.* **25**, 659 (1974).
14. I. Adachi, pers. comm. 1976.
15. K. Iga, *Appl. Opt.* **21**, 1024 (1980).
16. K. Iga and K. Yokomori, *Trans. IECE Jpn.* **58-C**, 283 (1975); K. Iga and N. Yamamoto. *Appl. Opt.* **16**, 1035 (1977).
17. T. Uchida, M. Furukawa, I, Kitano, K. Koizumi, and H. Matsumura, *IEEE J. Quant. Electron.* **QE-5**(6), 331 (June 1969).
18. K. Iga, S. Hata, Y. Kato and H. Fukuyo, *Jpn. J. Appl. Phys.* **13**(1), 79–86 (January 1974).
19. I. Kitano, K. Koizumi, H. Matsumura, T. Uchida, and M. Furukawa, *Suppl. J. Jpn. Soc Appl. Phys.* **39**, 63–70 (1970).
20. Y. Ohtsuka, *Appl. Phys. Lett.* **23**(5), 247–248 (September 1973).
21. K. Iga, K. Yokomori, and T. Sakayori, *Appl. Phys. Lett.* **26**(10), 578–579 (May 1975).
22. K. Iga and K. Yokomori, *Trans. IECE Jpn.* **58-C**(5), 283–285 (May 1975).
23. W. E. Martin, *Appl. Opt.* **14**(10), 2427–2431 (October 1975).
24. K. Nishizawa pers. comm. 1976.
25. K. Iga, N. Yamamoto and Y. Matsuura, *Trans. IECC Jpn.* **E60**, 239 (1977).
26. T. Miyazawa, K. Okada, T. Kubo, K. Nishizawa, I. Kitano, and K. Iga, *Appl. Opt.* **19** 1113 (1980).
27. K. Iga, *Appl. Opt.* **19**, 1939 (1981).
28. M. Saruwatari and T. Sugie, *Electron. Lett.* **16**, 955 (1980).
29. J. Crank, "The Mathematics of Diffusion." Clarendon Press, Oxford, 1975.
30. M. Born and E. Wolf, "Principles of Optics," 5th ed. Pergamon Press, Oxford, 1975.
31. T. Imamura, "Physics and Green's Function (in Japanese). Iwanami, Tokyo, 1978.
32. E. W. Marchand, "Gradient Index Optics." Academic Press, New York, 1978.
33. Y. Ohtsuka, *Appl. Phys. Lett.* **23**, 247 (1973).

34. K. Iga, K. Yokomori, and T. Sakayori, *Appl. Phys. Lett.* **26**, 578 (1975).
35. K. Iga and N. Yamamoto, *Appl. Opt.* **16**, 1305 (1977).
36. I. Kitano, K. Koizumi, H. Matsumura, T. Uchida, and M. Furukawa, "Proc. 1st Conf. Solid State Devices, Tokyo, 1969, *Oyo Buturi J. Jpn. Soc. Appl. Phys.* Suppl. **39**, 63 (1970).
37. H. Kawanishi and Y. Suematsu, *Trans. Inst, Electron. Commun. Eng. Jpn.* **E60**, 231 (1977).
38. J. Koyama and H. Nishihara, *Kouha Denshi Kougaku Opto-Electronics Engineering* (in Japanese). Corona Publishing, Tokyo, 1978.
39. T. Izawa and H. Nakagome, *Appl. Phys. Lett.* **21**. 584 (1972).
40. S. P. Mukherjee, Topical Meeting on Gradient-Index Optical Imaging Systems, Hawaii, TuAl, (1981).
41. I. Kitano, K. Koizumi, and H. Matsumura, *J. Jpn. Soc. Appl. Phys.*, Suppl. **39**, 63 (1970).
42. T. Izawa and H. Nakagome, *Appl. Phys. Lett.* **21**, 584 (1972).
43. P. B. Macedo and T. A. Litovitz, U.S. Patent, 3, 938, 974 (1976)
44. A. Panafieu *et al, Phys. Chem. Glasses* **21**, 22 (1980).
45. F. P. Kapron, D. B. Keck, and R. D. Maurer, *Appl. Phys. Lett.* **17**, 423 (1970).
46. W. G. French, A. Pearson, G. W. Tasker, and J. B. MacChesney, *Appl. Phys. Lett.* **23**, 338 (1973).
47. T. Izawa, S. Kobayashi, S. Sudo, and F. Hanawa, *IOOC'77* Cl-1 (1977).
48. G. D. Khoe, H. G. Kock, J. A. Luijendijk, C. H. J. van den Brekel, and D. Kuppers. *ECOC'81* 7.6 (1981).
49. Y. Ohtsuka Appl. Phys. L ett. **23**, 247 (1973).
50. M. Oikawa, K. Iga, and T. Sanada, *Jpn. J. Appl. Phys.* **20**, L51 (1981).
51. Y. Koike, Y. Kimoto and Y. Ohtsuka,"Top. Meet., Gradient-Index Opt. Imaging Systems," TuBl (1981).
52. J. J. Miceli, Jr., Topical Meeting on Gradient-Index Optical Imaging Systems, Hawaii, TuA2 (1981).

CHAPTER 8

Measurement of Index Distributions

8.1 OPENING REMARKS

As mentioned in the preceding sections, the refractive index profile plays an important role in optical device characterization. In particular, the transmission capacity of distributed-index multimode fibers greatly depends on index profile, and so it is necessary to closely control it to obtain a large transmission capacity.

Transmission characteristics of the nearly parabolic index fiber, such as optimal index profile, theoretical transmission capacity limit, index profile tolerance, and effect of material dispersion, have been studied.[1-10] From these analyses, it has become clear that the index profile tolerance in the parabolic index fiber is very limited, as shown in Fig. 8.1-1. It is necessary to control the index profile precisely to obtain the desired very large transmission capacity, i.e., 10 GHz km. Consequently, it becomes necessary to be able to measure the refractive index profile accurately during production to control it.

Correspondingly, the cutoff frequency of single-mode fibers also depends strongly on the index profile, as shown in Fig. 8.1-2. Although a single-mode fiber is designed to have a uniform index profile in the core, the actual drawn fiber has some amount of index gradient in the core. The cutoff wavelength of a single-mode fiber can be designed from the index profile of its preform rod if its index profile is similar to that of the preform rod. Therefore, it becomes necessary to measure the index profile of the preform rod to obtain the cutoff wavelength of a single-mode fiber. Furthermore, it is necessary to establish an index profile measurement technique for preform rods for the testing and sorting needed to manufacture well-controlled graded-index multimode fibers.

Many microlenses are utilized in light-wave components. Aberration of the microlenses reduces the performance of the components, and measuring methods are necessary for their characterization. A direct measurement of

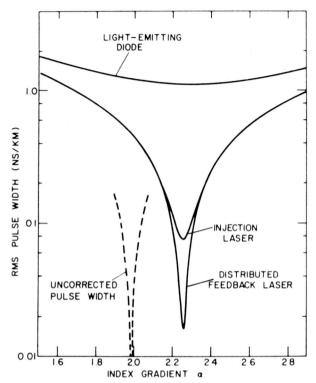

FIGURE 8.1-1.   Pulse broadening versus index gradient a of mulimode traded index fiber. (After R. Olshansky and D. B. Keck.[8])

index distribution in DI microlenses is effective for doing this, and the result can be fed back to the fabrication processes. Other methods of measuring aberrations are introduced in Chapter 9.

8.2 INDEX-PROFILING METHODS

Since the advent of graded-index fibers, many refractive-index profile-measuring methods have been studied. Early some methods were applicable to only step index fibers or only to parabolic index fibers. In 1969 a method applicable to an arbitrary index profile was reported.[11]

Since 1973 a number of primary measurement methods have been reported.[12–15] These are classified as destructive or nondestructive methods. The destructive method requires much time to prepare samples with very flat end surfaces. Although the nondestructive method does not require such a troublesome process, it does require computer calculation to obtain a

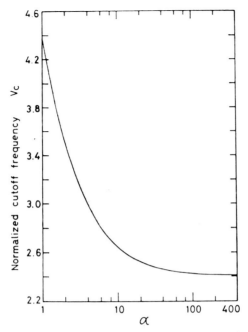

FIGURE 8.1-2.   Cutoff frequency versus index gradient a of single mode fiber.

refractive-index profile from a scattering pattern, or an interference pattern, etc. In this section, measuring methods for arbitrary refractive-index profiles are summarized.

A. DESTRUCTIVE METHODS

Longitudinal Interference Methods   The fiber sample (or preform rod sample) is cut in a thin round slice and its surfaces are polished to be optically flat. This thin sample is examined under an inteference microscope as shown in Fig. 8.2-1. The refractive index profile is calculated from fringe shift. An automatic measuring system has been developed by incorporating an image processing apparatus, such as a vidicon camera, and a computer.[16,17] The spatial resolution limit is about 0.7 μm.[18] If the sliced sample is too thick, the incident ray is refracted through the sample and causes an error.[19] Therefore, the sample must usually be polished to less than 100 μm in thickness. This takes much time, which deters this method from being introduced into the fiber manufacturing process as a testing system. Accuracy is limited to about 0.0005 because of the roughness of the polished surfaces.[16]

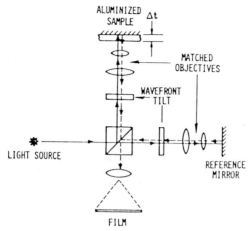

FIGURE 8.2-1.   Measuring setup for the longitudinal interference method. (After W. E. Martin.[13])

Reflection Method   The reflection coefficient of a dielectric material is related to the refractive index at the incident surface.[20,21] The refractive-index profile of an optical fiber can be measured by utilizing this principle. A laser light beam with a small spot size is focused into the end surface of an optical fiber, and the reflection coefficient is measured by comparing the incident and reflected light intensity as shown in Fig. 8.2-2. The refractive-index profile is obtained from the reflection coefficient profile by shifting the reference point. Accuracy is strongly affected by the flatness of the end surface. A fractured end surface gives better results than a polished end surface.[22] For borosilicate fibers, the result changes rapidly with time[22] because of atmospheric exposure of the dopant. Spatial resolution is usually limited to about 1–2 μm by the spot size of the incident beam. This effect of finite-beam spot size can be

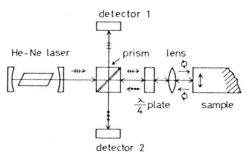

FIGURE 8.2-2.   Measuring setup for the reflection method. (After M. Ikeda, M. Tateda and H. Yoshikiyo.[21])

corrected by numerical calculation.[23] A 0.3 μm spatial resolution and 5% total accuracy of the refractive index has been obtained by this correction.[23]

Near-Field Pattern Method   When all the guided modes are excited uniformly by using an incoherent light source, the near-field pattern of output optical power is similar to the refractive-index profile.[24] Since it is difficult to strictly satisfy the incident condition and the near-field pattern is affected by leaky modes and the absorption loss difference of guided modes, this method cannot provide accurate measurements. Although several improvements, such as a correction factor[25-28] for leaky modes, a refracting ray method[29-31] that is free from leaky mode effect, and a spot scanning method[27,28,32] have been made to improve accuracy, this method is being used only as an auxiliary technique.

XMA Method   The XMA method measures the dopant concentration profile, which relates to the index profile, by means of an XMA (x-ray micro analyzer).[33,34] The contribution of dopants such as P_2O_5, GeO_2, and B_2O_5 to the refractive index can be obtained separately, but accuracy is not good because of low S-to-N ratio.

SEM Method (Etching Method)   When the end surface of an optical fiber is chemically etched, the etching speed depends on the dopant concentration. Therefore, the unevenness of the etched surface represents the dopant concentration profile. This unevenness can be observed by an SEM (scanning electron-beam microscope).[35-37]

Far-Field Pattern Method   This method utilizes the far-field pattern of output optical power instead of the near-field pattern.[38] The method of Ref. (39) is applicable to only single mode fibers. The former method does not give good accuracy because of modal interference within the far-field.[38] The latter requires an optical detector with a large dynamic range and the error is more than 5%.[39]

B. Nondestructive Methods

Scattering Pattern Method   The scattering pattern method is classified into both a forward scattering pattern method[40-42] and a backward scattering pattern method.[43-45] In the case of the forward scattering pattern method,[40] the sample is immersed in an index-matching oil and the forward scattering pattern is observed by using the apparatus shown in Fig. 8.2-3. The refractive-index profile is calculated from the scattering pattern by using a computer.[40,41] The error of this method is determined by the product of the core

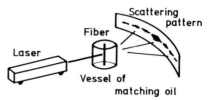

FIGURE 8.2-3.   Optical setup for measuring the forward scattering pattern. (After T. Okoshi and K. Hotate.[40])

radius a and the index difference Δn, and it increases with an increase of $a \, \Delta n$. As a numerical example, when $a \, \Delta n = 0.04$ mm, the error is 5%. Therefore, this method is applicable to only single-mode fibers. Since this method requires many sampling points (500–1000), it is necessary to collect the data automatically.[40]

On the other hand, the index profile can also be obtained from the backward scattering pattern.[43–45] This method does not require index-matching oil and is applicable to thick samples such as preformed rods.[42] However, since the backward scattering pattern is stained with externally reflected light, it is not suited for precise measurements.[45] Furthermore, the accuracy of this method is very sensitive to the elliptical deformation of the core cross section.[44,45]

Transverse Interference Method   The fiber sample is immersed in index-matching oil and is observed in its transverse direction by using an interference microscope.[46–52] The index profile is calculated from fringe shift.[47,48] Before the author began this study,[48] analysis based on straight-ray trajectory[47,51] had always been used to calculate index profile. However, it is now known that accuracy can be increased by using an analysis that includes ray refraction.[48] There also exists another method[52] that utilizes ray refraction angle to calculate the index profile, but the accuracy is not very good. The details of this method are discussed in the following section.

Transverse Differential Interference Method   The transverse differential method is an interference method modified to apply to thick samples, such as focusing rod lenses[53–55] and optical-fiber perform rods.[56] Instead of a transverse interference pattern, a transverse differential interference pattern, differentiated with respect to the transverse distance, is used to calculate the index profile. The details of this method are also discussed in the following section.

Focusing Method   When an optical fiber with an axially symmetric index distribution is illuminated in its transverse direction by an incoherent light source, the fiber acts as a lens so that the incident light is focused on a plane placed behind the fiber. If the light intensity distribution is uniform with respect to the incident place, the index profile can be calculated from the focused light intensity distribution.[57] This method can be applied to preform rods[58,59] as well as fibers[60] and is one of the promising methods, along with the transverse and transverse differential interference methods. This method can be applied to axially nonsymmetric preforms.[61]

X-Ray Absorption Profile Method A transmittivity pattern is obtained when a preform rod is irradiated with x-rays in its transverse direction.[62] Since the absorption coefficient of germanium is relatively large, the germanium concentration profile can be calculated from this transmittivity pattern by means of an Abelian inversion.

Spatial Filtering Method The spatial filtering method is similar to the focusing method. However, the deflection angle is detected and directly

TABLE 8.2-I

VARIOUS MEASURING METHODS OF INDEX PROFILE

		Measurement time	Accuracy	Sample preparation	Correction of elliptical deformation
Longitudinal interference method		long	good	mirror polish	easy
Near-field pattern method		short	fairly good	cleave	possible
Reflection method		medium	fairly good	cleave	possible
Scattering pattern method		short	fairly good	not necessary	difficult
Transverse interference method		short	good	not necessary	possible
Focusing method		short	good	not necessary	not practical
Spatial filtering method		short	good	not necessary	possible

displayed by an optical spatial filter,[63] which reduces detection error. This method can be applied to axially nonsymmetrical preforms.[64]

As mentioned above, many index profile measurement methods have been proposed. Each method has merits and demerits, and the features of these methods are summarized in Table 8.2-I. The accuracy of the transverse interference and transverse differential interference methods will be discussed in the following section.

8.3 SHEARING INTERFERENCE METHOD

A. PRINCIPLES

The transverse interference method is similar to the tomography methods in the fields of aerodynamics,[65,66] seismology,[67] plasma physics and astrophysics.[68] In these methods, the phase shift of a probing wave such as a light wave or an earthquake wave is detected and used to calculate the axially symmetric internal structure of a phase object. In the transverse interference method, an optical fiber or perform rod sample is regarded as a phase disturbance which has an axially symmetric index profile, and the phase shift of a probing light is detected as fringe shift by using a formula which is the Abelian inversion of an Abelian integral equation. Thus far, in calculating the index profile from the fringe shift, the probing beam incident on the sample has been assumed to traverse a straight path through the sample, and so it has been necessary to derive an exact formula which corrects for the probing ray refraction. Hunter and Schreiber[69] made this trial, but they committed an error in that the tangent term in the fringe shift theory[70] was omitted.

In the following, an exact formula which corrects for ray refraction is derived, and the accuracy of this formula is investigated by some computer simulations.

It is assumed that an optical fiber or a preform rod sample is immersed in index-matching oil whose refractive index is close to that of the sample's cladding. The sample is observed in its transverse direction by using an interference microscope as shown in Fig. 8.3-1. The probing beam, after having traversed the core, is divided into two arms by a Mach–Zehnder interferometer, and the beam W_2 is shifted by a shearing prism.

In the first case, when the shearing distance is made larger than the size of the core, as shown in Fig. 8.3-2, the flat wavefront of one of the divided beams interferes with the retarded wavefront of the other, which has traversed the core. Thus an ordinary interference pattern (total shearing pattern) is observed as shown in Fig. 8.3-3. The measured value of the fringe shift $R(y)$ represents the transverse optical pathlength difference of the core from the outer cladding or the matching oil.

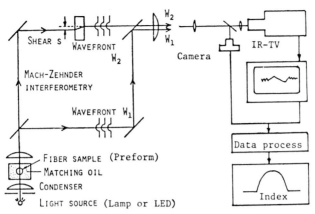

FIGURE 8.3-1.   Optical setup for transverse interferometry (After Y. Kokubun and K. Iga.[49,50,56])

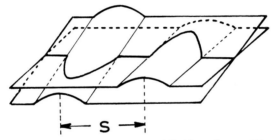

FIGURE 8.3-2.   Interference of two wavefronts shifted by a distance S (total shearing).

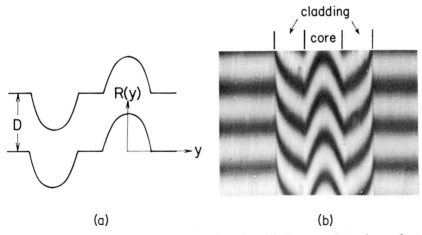

(a)   (b)

FIGURE 8.3-3.   Interference fringe of Figure 8.3-2. (a) Cross section of wavefront.
(b) Photograph of actual interference pattern.

On the other hand, when the shearing distance becomes small enough compared with the size of the core, the observed fringe shift approximately represents the differential of optical pathlength difference with respect to the transverse distance y, as shown in Fig. 8.3-4. The fringe shift $R'(y)$ is expressed in terms of $R(y)$ and the shearing distance s by

$$R'(y) \simeq S \, dR(y)/dy. \qquad (8.3\text{-}1)$$

The refractive-index profile can be obtained from the measured values of $R(y)$ or $R'(y)$ with the aid of analytic formulas as stated in the next section.

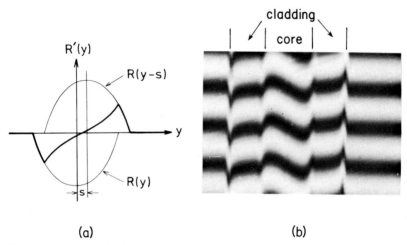

(a)   (b)

FIGURE 8.3-4.   Principle of the differential interference fringe. (a) Cross section of wavefront. (b) Photograph of actual differential interference pattern.

When the core radius is smaller than several hundred micrometers, ordinary (total shearing) interferometry is more suitable for measurement, because in differential interferometry the shearing distance S cannot be made small enough compared with the core radius to obtain an appreciable fringe shift. On the other hand, when the core radius is larger than 1 mm, as in the case of preform rods and focusing rod lenses, differential iterferometry is more suitable for observation, because in ordinary interferometry the fringe shift overflows from the field of microscope due to the large optical path difference.

In the following subsections we refer to the transverse interference method as "TI method" and the transverse differential interference method as "TDI method."

The interference pattern was photographed or observed by an IR–ITV camera. The fringe shift was read from the photograph or the picture displayed on TV monitor. The refractive-index profile was calculated from the fringe shift data with the aid of the formulas derived in the following section.

B. Formulas for Calculating Refractive-Index Profile

A test beam which consists of parallel light rays illuminates a sample in the transverse direction. The probing ray is refracted through the core, which has an axially symmetric refractive-index profile, as shown in Fig. 8.3-5. When the y axis is located on the lens–object plane, an incident ray y distance from the x axis is focused at $y' = y \sec \psi$ in the image plane. The phase shift of the probing ray as a function of y' is obtained by observing the fringe shift in the interference pattern.

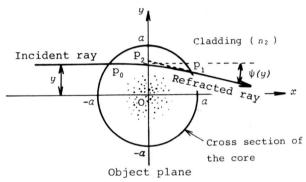

FIGURE 8.3-5.   Ray trajectory in a core which has an axially symmetric refractive index profile [The angle $\psi(y)$ is on the order of 10^{-2} rad in magnitude] (After Y. Kokubun and K. Iga.[49])

The fringe shift $R(y \sec \psi)$ in the lens–image plane was given in Ref. 70 as follows:

$$\frac{\lambda}{D} R(y \sec \psi) = \int_{P_0}^{P_1} n(r)\, ds - 2n_2\sqrt{a^2 - y^2} - n_2 y \tan \psi, \quad (8.3\text{-}2)$$

where a is the core radius, n_2 the refractive index of the cladding, D the distance between interference fringes corresponding to one wavelength, and λ the wavelength of the light source. The optical pathlength from P_0 to P_1 and the refraction angle $\psi(y)$ are expressed[69] in terms of the local refractive index $n(r)$ and the transformed argument $u = rn(r)$ by

$$\int_{P_0}^{P_1} n(r)\, ds = 2n_2\sqrt{a^2 - y^2} - 2 \int_{n_2 y}^{u_2} \left\{ \frac{d \ln n(u)}{du} \right\} \frac{u^2\, du}{\sqrt{u^2 - n_2^2 y^2}}, \quad (8.3\text{-}3)$$

$$\psi(y) = -2n_2 y \int_{n_2 y}^{u_2} \left\{ \frac{d \ln n(u)}{du} \right\} \frac{du}{\sqrt{u^2 - n_2^2 y^2}}, \quad (8.3\text{-}4)$$

respectively, where $u_2 = n_2 a$.

Therefore, substitution of Eqs. (8.3-3) and (8.3-4) into Eq. (8.3-2) gives an integral equation, but this is difficult to solve analytically because it includes an integral in the argument of the trigonometeric function. Hunter and Schreiber[69] ignored the tan ψ term in Eq. (8.3-2), but this term is of the same order of magnitude as the optical pathlength term, so they committed a serious omission.

The trigonometric functions in Eq. (8.3-2) can be expressed in terms of a Taylor series expansion:

$$\tan \psi(y) = \sum_{j=1}^{\infty} \frac{2^{2j}(2^{2j} - 1)B_j}{(2j)!} \psi^{2j-1}(y) \qquad \left(|\psi| < \frac{\pi}{2} \right)$$

$$= \psi(y) + \tfrac{1}{3} \psi^3(y) + \tfrac{2}{15} \psi^5(y) + \cdots$$

and

$$\sec \psi(y) = 1 + \sum_{j=2}^{\infty} \frac{E_j}{(2j)!} \psi^{2j}(y) \qquad \left(|\psi| < \frac{\pi}{2} \right)$$

$$= 1 + \tfrac{1}{2} \psi^2(y) + \tfrac{5}{24} \psi^4(y) + \cdots,$$

where B_j and E_j are Bernoulli's and Euler's numbers, respectively.

In the first-order approximation, $\tan \psi$ and $\sec \psi$ in Eq. (8.3-2) are approximated by their first terms in a Taylor series expansion. This simplification gives an Abelian integral equation as follows:

$$2 \int_{n_2 y}^{u_2} \ln n(u) \frac{u \, du}{\sqrt{u^2 - n_2^2 y^2}} = \frac{\lambda}{D} R(y) + 2\sqrt{u_2^2 - n_2^2 y^2} \ln n_2, \quad (8.3\text{-}5)$$

and its solution is given by

$$n(u) = n_2 \exp\left\{ -\frac{1}{\pi} \int_{u/n_2}^{a} \left[\frac{\lambda}{D} \frac{dR(y)}{dy} \right] \frac{dy}{\sqrt{n_2^2 y^2 - u^2}} \right\}. \qquad (8.3\text{-}6)$$

The index profile, as a function of radius r, is explicitly expressed with the aid of Eq. (8.3-6) and the relation $r = u/n(u)$. Since the denominator of the integrand in Eq. (8.3-6) reaches zero when $y = u/n_2$, data reduction is performed with the aid of

$$n(u) = n_2 \exp\left[-\frac{1}{\pi} \int_0^{\pi/2} \frac{\lambda}{D} \frac{dR(y)}{dy} \frac{\sqrt{a^2 - (u/n_2)^2} \cos \theta}{n_2 \sqrt{(u/n_2)^2 + [a^2 - (u/n_2)^2] \sin^2 \theta}} \, d\theta \right],$$

$$(8.3\text{-}7)$$

after the variable is changed by

$$y = \sqrt{(u/n_2)^2 + [a^2 - (u/n_2)^2] \sin^2 \theta}. \qquad (8.3\text{-}8)$$

This is the formula which corrects for ray refraction in the first-order approximation.

When the probing rays are assumed to traverse straight through the sample, the data reduction formula is known to be

$$n(r) = n_2 - \frac{1}{\pi} \int_r^a \left\{ \frac{\lambda}{D} \frac{dR(y)}{dy} \right\} \frac{dy}{\sqrt{y^2 - r^2}}. \qquad (8.3-9)$$

This corresponds to the case when the right-hand side of Eq. (8.3-6) is approximated by the first two terms of a Taylor series expansion of the exponential function and u/n_2 is approximated by r.

In the second-order approximation, the refraction angle $\psi(y)$ is evaluated by substituting the result of the first-order approximation into Eq. (8.3-4), and the second term in a Taylor series expansion of tan ψ is treated as a correction term. After some simplification, substitution of Eq. (8.3-6) into Eq. (8.3-4) gives

$$\psi_2(y) = -\frac{\lambda}{n_2 D} \frac{dR(y)}{dy}. \qquad (8.3-10)$$

When the corrected fringe shift in the second-order approximation is defined by

$$F_2(y) = \frac{\lambda}{D} R \left\{ y \left[1 + \frac{\psi_2^2(y)}{2} \right] \right\} + \frac{n_2 y}{3} \psi_2^3(y), \qquad (8.3-11)$$

the refractive-index profile in the second-order approximation is given by

$$n(u) = n_2 \exp\left(-\frac{1}{\pi} \int_{u/n_2}^a \frac{dF_2(y)}{dy} \frac{dy}{\sqrt{n_2^2 y^2 - u^2}} \right), \qquad (8.3-12)$$

Higher-order approximations can also be performed, resulting in the following formulas:

$$\psi_i(y) = -\frac{1}{n_2} \frac{dF_{i-1}(y)}{dy}, \qquad (8.3-13)$$

$$F_i(y) = \frac{\lambda}{D} R \left\{ y \left[1 + \sum_{j=1}^{i-1} \frac{E_j}{(2j)!} \psi_i(y)^{2j} \right] \right\}$$

$$+ n_2 y \sum_{j=2}^i \frac{2^{2j}(2^{2j}-1)B_j}{(2j)!} \psi_i(y)^{2j-1}, \qquad (8.3-14)$$

and

$$n(u) = n_2 \exp\left(-\frac{1}{\pi} \int_{u/n_2}^a \frac{dF_i(y)}{dy} \frac{dy}{\sqrt{n_2^2 y^2 - u^2}} \right), \qquad (8.3-15)$$

where i expresses the order of successive approximation. Initial conditions are $\psi_1(y) = 0$ and $F_1(y) = \lambda R(y)/D$.

This higher-order approximation is necessary when the index difference between the core center and the cladding is rather large or there is a steep index gradient locally in the core.

C. ACCURACY OF DATA REDUCTION FORMULAS

Computer Simulation   To investigate the accuracy of the data reduction formula Eq. (8.3-6) and Eqs. (8.3-10)–(8.3-12), it is assumed that an index profile

$$n(r) = \begin{cases} n_1\sqrt{1 - 2\Delta(r/a)^\alpha}, & r \le a, \\ n_2 & r > a, \end{cases} \qquad (8.3\text{-}16)$$

where $\Delta = (n_1 - n_2)/n_2$, has been given, the transverse interference fringe shift $R(y)$ has been numerically calculated from the given profile, and the index profile has been recalculated numerically from the fringe shift with the aid of Eq. (8.3-6) for the first-order approximation and Eqs. (8.3-10)–(8.3-12) for the second-order approximation. In these computer simulations, the error in calculating the fringe shift is small compared with the error of numerical calculation of the index profile.

The error ε is defined by

$$\varepsilon = \frac{n_{cal}(r) - n(r)}{n_1},$$

where $n_{cal}(r)$, $n(r)$, and n_1 are the refractive indices, calculated with the aid of Eq. (8.3-6) and Eqs. (8.3-10)–(8.3-12), given initially by Eq. (8.3-16), and of the core center, respectively. Since the error ε of the first-order approximation is expected to be nearly proportional to Δ^3 and considering the fact that the refraction angle ψ is proportional to $\Delta[= (n_1 - n_2)/n_1]$, it is natural to normalize by Δ^3.

Figure 8.3-6 illustrates the error of the previous analysis (dashed curve), of the first-order approximation (dashed–dotted curve), and of the second-order approximation (solid curve), respectively. From this figure, the error of the second-order approximation is about 10^{-3} times smaller than that of the previous analysis.

Figure 8.3-7 shows the error for different values of Δ. From this figure, it can be observed that the error of the first-order approximation is nearly proportional to Δ^3. This is thought to be natural when we consider Eq. (8.3-11) which expresses $F_2(y)$ as related to ψ^3 and when $\psi \propto \Delta$ is taken into account. The error of the second-order approximation remains small even when Δ takes a rather large value, i.e., 0.05.

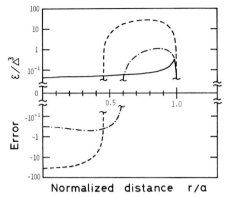

FIGURE 8.3-6.   Principle errors obtained by computer simulations for $\alpha = 2$ and $\Delta = 0.015$. Errors are shown for the previous analysis (———), the first-order approximation (—·—·), and the second-order approximation (———). (After Y. Kokubun and K. Iga.[50])

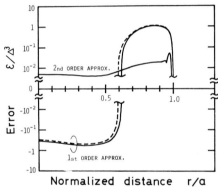

FIGURE 8.3-7.   Principle errors as Δ takes the values 0.015 (———) and 0.05 (———) for $\alpha = 2$. (After Y. Kokubun and K. Iga.[50])

Figure 8.3-8 shows the error for different values of the exponent α in Eq. (8.3-16). When the value of α is large, the error of the first-order approximation becomes rather large. However, the error of the second-order approximation is smaller than that of the first-order approximation by two orders of magnitude even when $\alpha = 10$.

D.   OTHER ERROR FACTORS AND THEIR CORRECTIONS

If the refractive index of the index-matching oil n_{oil} is not equal to that of the cladding n_2, the error caused by the index mismatch is maximum at the core center. This maximum error δn_1 is proportional to the index difference $n_2 - n_{\mathrm{oil}}$ and depends on the ratio of the core radius a and the fiber radius R_{f}.

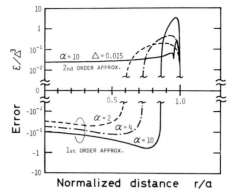

FIGURE 8.3-8.   Principle errors as α takes various values. (After Y. Kokubun and K. Iga.[50])

Figure 8.3-9 shows the maximum error δn_1 normalized by the index mismatch $n_2 - n_{\text{oil}}$ versus the ratio of core and fiber radii. The dashed curve represents the error without correction, and the solid curve the error with a correction which is expressed by

$$a_{\text{real}} = \frac{n_{\text{oil}}}{n_2} a_{\text{meas}}, \qquad (8.3\text{-}17)$$

$$R_{\text{cor}}(y) = R(y) - (n_2 - n_{\text{oil}}) \sqrt{R_{\text{f}}^2 - y^2}. \qquad (8.3\text{-}18)$$

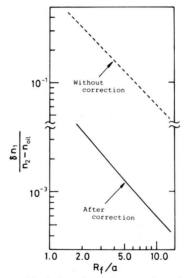

FIGURE 8.3-9.   Error caused by index mismatch versus the ratio of core and fiber radii.

The refractive-index difference $n_2 - n_{oil}$ is determined from the fringe shift of the cladding region on the basis of the matching oil. If the refractive index of the matching oil is previously known, the index of the cladding can be determined on the basis of that of the matching oil. However, since the refractive index of fluids generally depends on temperature and wavelength, the temperature coefficients of several kinds of matching oil was previously measured at wavelengths of 0.589 and 0.867 μm, respectively.

As an other error factor, the spectral width of the light source should be considered. However, the spectral width acts as a factor which increases the reading error of the fringe shift, and it will be discussed in a later section as the reading error of the fringe shift.

E. SPATIAL RESOLUTION

In the preceding section, the error in measuring refractive index was discussed. However, spatial resolution is another important factor of accuracy. Spatial resolution is determined by both the resolving power of the microscope used and the maximum spatial frequency limited by the sampling interval of the data. The resolving power d is approximately expressed by

$$d = 0.61\lambda/\text{NA},$$

where λ is the light source wavelength and NA the numerical aperture of the microscope objective lens. The resolving power is on the same order of magnitude as the wavelength.

The maximum spatial frequency limited by the sampling interval is calculated in the following way. The Fourier transform of the data reduction formula (8.3-9) gives

$$N(\Omega) = \int_0^\infty g(y) J_0(\Omega y) y \, dy \qquad (g(y) = 0, \quad a < y), \qquad (8.3\text{-}19)$$

where $N(\Omega)$ is the Fourier transform of $n(r) - n_2$,

$$g(y) = -(\lambda/4D)(1/y)(dR/dy),$$

and Ω is the spatial angular frequency. Equation (8.3-19) states that $N(\Omega)$ is the Hankel transform of $g(y)$. When $N(\Omega) = 0$ for $\Omega > \Omega_m$, the Hankel inversion of Eq. (8.3-19) gives

$$g(y) = \Omega_m^2 \int_0^1 N(k) J_0(\Omega_m ky)k \, dk. \qquad (8.3\text{-}20)$$

On the other hand, $N(\Omega)$ is expressed by the Fourier–Bessel transform as

$$N(k) = \sum_{v=1}^{\infty} a_v J_0(\alpha_{0v}k), \qquad (8.3\text{-}21)$$

where α_{0v} is the vth zero of $J_0(x)$ and

$$a_v = \frac{2}{J_1^2(\alpha_{0v})} \int_0^1 N(t) J_0(\alpha_{0v}t)t \, dt. \qquad (8.3\text{-}22)$$

Substitution of Eq. (8.3-20) into Eq. (8.3-22) gives

$$a_v = \frac{2}{J_1^2(\alpha_{0v})\Omega_m^2} g\left(\frac{\alpha_{0v}}{\Omega_m}\right). \qquad (8.3\text{-}23)$$

Further, substituting Eq. (8.3-23) into Eq. (8.3-21), the following sampling formula for $N(k)$ is obtained:

$$N(k) = \sum_{v=1}^{\infty} \frac{2}{J_1^2(\alpha_{0v})\Omega_m^2} J_0(\alpha_{0v}k) g\left(\frac{\alpha_{0v}}{\Omega_m}\right) \qquad (8.3\text{-}24)$$

Since $g(y) = 0$ for $y > a$, the maximum value Ω_m is approximated by

$$\Omega_m = \alpha_{0N}/a, \qquad (8.3\text{-}25)$$

where N is the number of data samples. When N is much larger than unity, α_{0N} is approximated by

$$\alpha_{0N} \simeq \pi(4N-1)/4,$$

and so Ω_m is rewritten as

$$\Omega_m = \frac{\pi}{a}\frac{4N-1}{4} \simeq \frac{\pi}{a}N. \qquad (8.3\text{-}26)$$

From Eq. (8.3-26), it is seen that the maximum spatial frequency is inverse to the sampling interval a/N. Since the number of sampled data is usually ~ 50–100, the sampling interval is smaller than or on the same order of magnitude as the resolving power of the microscope, in the case of optical fibers whose core radius is several micrometers. On the other hand, in the case of preformed rods, the spatial resolution is nearly equal to the sampling interval.

F. REFRACTIVE-INDEX PROFILE MEASUREMENT OF GRADED-INDEX MULTIMODE FIBERS

Figure 8.3-10 shows an ordinary transverse interference (total shearing) pattern of a silica optical fiber made using the CVD method. Figure 8.3-11 shows a transverse interference pattern of a graded-index optical fiber made of compound glass. In Figs. 8.3-10 and 8.3-11, a tungsten lamp was used as a light source and a red filter was placed in front of the light source to obtain sharp. fringes.

Figure 8.3-12 shows the refractive-index profile of a silica fiber calculated from the interference pattern shown in Fig. 8.3-10. Diameters of core and

|← core →|

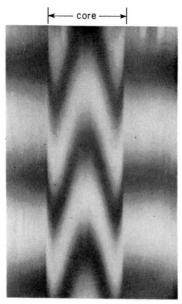

FIGURE 8.3-10. Transverse interference pattern of a silica optical fiber made by using the CVD method. (After Y. Kokubun and K. Iga.[49])

|← core →|

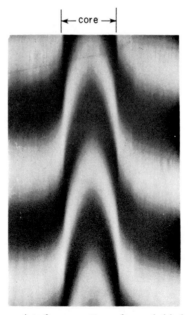

FIGURE 8.3-11. Transverse interference pattern of a graded-index fiber made of compound glass. (After Y. Kokubun and K. Iga.[50])

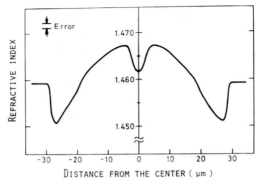

FIGURE 8.3-12.   Refractive index profile of a silica glass fiber, where $\lambda = 0.67\ \mu$m, $n_2 = 1.4593$, and $a = 29.6\ \mu$m. (After Y. Kokubun and K. Iga.[49])

cladding are 59.2 and 125 μm, respectively. The refractive index of the cladding was determined to be 1.4593 by using the fringe shift from the matching oil whose refractive index is 1.4598. The core radius was divided into 70 equal parts, and the sampled fringe shift corresponding to each part was used to calculate the refractive-index profile with the aids of Eqs. (8.3-10)–(8.3-12). In this measurement, error due to the index difference between the cladding and the matching oil was corrected for with the aid of Eqs. (8.3-17) and (8.3-18).

The index difference between the maximum and the minimum values of refractive index was 0.0159 and the least-squares fit of the exponent α of the α-power law index profile Eq. (8.3-16) was calculated to be 2.53.

Figure 8.3-13 shows the refractive-index profile of a compound glass fiber calculated from the transverse interference pattern shown in Fig. 8.3-11.

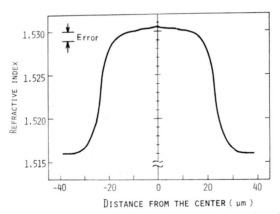

FIGURE 8.3-13.   Refractive index profile of a compound glass fiber, where $\lambda = 0.67\ \mu$m and $n_2 = 1.5160$. (After Y. Kokubun and K. Iga.[50])

Diameters of core and cladding are 69 and 200 μm, respectively. The refractive index of the cladding was determined to be 1.5160 by comparing it with that of the matching oil. The number of data samples was 95. The least-squares fit of the exponent α was about 4.0.

In these measurements, the principal error is very small, as mentioned in the previous section. However, there remains an error resulting from the data reading at the fringe center which must be considered as a factor influencing total accuracy. Therefore, the error resulting from the data reading will be investigated in the following section.

G. Evaluation of Reading Error

The effect of the reading error generated in determining the fringe center was investigated by computer simulation. When the difference between the indices initially given and those calculated is designated δn, its standard deviation is defined by

$$\varepsilon = \sqrt{\langle \delta n^2 \rangle - \langle \delta n \rangle^2}. \tag{8.3-27}$$

Now, designate the reading error δR. A computer simulation was performed to make δR clear, as follows: (1) The initial fringe shift of the α-power law index profile was calculated numerically. (2) The reading error was assumed to be governed by the Gaussian distribution, and the quasi-Gaussian random number of the standard deviation of δR was added to the initially given fringe shift. (3) The refractive-index profile was calculated from this fringe shift, including quasi-Gaussian random noise, with the aid of Eqs. (8.3-10)–(8.3-12). (4) The standard deviation ε was calculated with the aid of Eq. (8.3-27), and ε was compared with δR for the various values of α in Eq. (8.3-16).

From this computer simulation, the relation between δR and ε for the transverse interference method was obtained as follows:

$$\frac{\varepsilon}{\Delta n} = K_t \frac{\lambda}{2a \, \Delta n} \frac{\delta R}{D}, \tag{8.3-28}$$

where K_t is the coefficient independent of the index profile and is expressed in terms of the sample number N as

$$K_t = 0.15(N - 1). \tag{8.3-29}$$

For example, in the case of $\Delta = 0.5\%$, $\lambda = 0.83$ μm, and $\delta R/D = 1.0\%$, the error resulting from the data reading is evaluated as shown in Fig. 8.3-14, from which it is seen that the error can be reduced by decreasing the number of data samples. But, the spatial resolution becomes worse with the decrease of the sample number if the spatial resolution due to data sampling becomes larger than the resolution power of the microscope, as mentioned in the

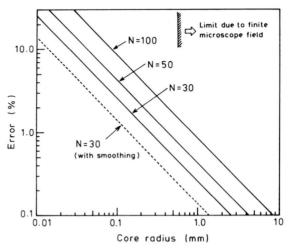

FIGURE 8.3-14.   Core radius dependence on the error resulting from data reading in the TI method, where $\Delta = 0.05\%$, $\lambda = 0.83$ μm, $\delta R/D = 1.0\%$, and N is the sampling number.

previous section. Therefore, the optimum value of the number of sampled data is given by

$$N \simeq a/\lambda. \qquad (8.3\text{-}30)$$

On the other hand, the error resulting from the reading error of the fringe center for the TDI method was evaluated in the same way. From the computer simulation, the relation between the reading error δR and ε defined by Eq. (8.3-27) for the TDI method was obtained as follows:

$$\frac{\varepsilon}{\Delta n} = K_d \frac{\lambda}{2s\,\Delta n} \frac{\delta R}{D}, \qquad (8.3\text{-}31)$$

where K_d is a constant number independent of the index profile and the sample number, $K_d = 0.19$. Equation (8.3-31) shows that the error resulting from the data reading is independent of the core radius. However, when the core radius becomes less than about 10 times the shearing distance, the fringe shift cannot be approximated by the differential of the optical pathlength difference between two paths passing through the core and the cladding, respectively. Therefore, the mean value of error increases with the decrease of core radius. Figure 8.3-15 shows the mean value of error versus the ratio of core radius to shearing distance. Since the total error is appraised by the sum of the mean value and the standard deviation, Figure 8.3-16 shows the sum of errors given in Fig. 8.3-15 and evaluated by Eq. (8.3-31) with an α value of 2. For comparison with the TI method, the error resulting from the data reading for the TI method is illustrated in the same figure.

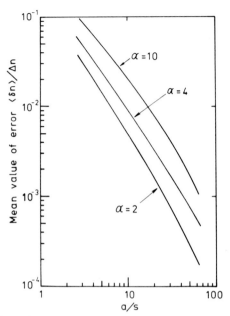

FIGURE 8.3-15.   Mean value of error versus the ratio of core radius to shearing distance. (After Y. Kokubun and K. Iga.[86])

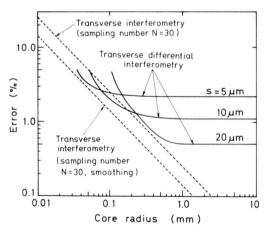

FIGURE 8.3-16.   Core-radius dependence of the error resulting from data reading in the TDI method where $\Delta = 0.5\%$, $\lambda = 0.83\ \mu m$, $\alpha = 2$, and $\delta R/D = 1.0\%$. (After Y. Kokubun and K. Iga.[56])

As mentioned in the previous sentence, the error for the TI method decreases with an increase in the core radius. But the fringe shift becomes larger with the increase in the core radius, and it finally overflows from the field of the microscope. Therefore, the TI method is limited to less than about 500 μm of the core radius. In contrast, the error for the TDI method is independent of the core radius as long as the shearing distance is small enough compared with the core radius. If a preform rod of $a > 1$ mm core radius is measured with a shearing distance of 20 μm, the total error is only about half of the reading error. Therefore, the TDI method is suitable for measurement of preform rods whose core radius is larger than 1 mm.

The optimum value of the shearing distance is obtained using the condition that the sum of the errors given by Eq. (8.3-31) and shown in Fig. 8.3-15 takes the minimum value as follows:

$$S \simeq \left(0.41 \frac{\lambda}{\Delta n} \frac{\delta n}{D} \right)^{0.4} a^{0.6}$$

$$\simeq 0.55 \times a^{0.6} \qquad\qquad (8.3\text{-}32)$$

H. REFRACTIVE-INDEX PROFILE MEASUREMENT OF OPTICAL FIBER PREFORMS AND DI RODS

As mentioned in Section 8.1, the refractive-index profile measurement of optical fiber preform rods is required in testing and sorting to obtain well-controlled graded-index multimode fibers and also to determine the cutoff frequency of single-mode fibers.[71-73] To measure the refractive-index profile of preform rods, methods applied to optical fibers such as the longitudinal interference method,[11-19] the near-field pattern method,[24,32,74] and the reflection method[20-23] seem to be applicable. However, since the preform rod is drawn into a fiber, its index profile must be measured nondestructively. The transverse interference method[46-52] can satisfy this demand with good accuracy. However, in applying this method to preform rods, the fringe shift becomes very large and overflows from the interference microscope field. An improvement in data collection[75,76] has been made to overcome this difficulty, but in the improved method the spatial resolution is limited because the amount of data for sampling the fringe shift is fixed.

In this section, the transverse differential interference method is applied to the nondestructive refractive-index profile measurement of preform rods. The measuring technique and index determination from the collected data are discussed.

Figure 8.3-17 shows the TDI pattern of a GeO_2-doped preform rod for a graded-index multimode fiber fabricated by the CVD method. The small

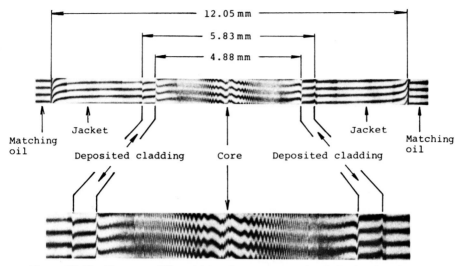

FIGURE 8.3-17.   Transverse differential interference pattern of a GeO_2-doped preform rod. (After Y. Kokubun and K. Iga.[56])

ripples in the fringe shift are due to the index fluctuation in each deposited layer. The number of layers was about 70. These ripples are magnified by the differential but have a rather small effect on the index profile. To determine the whole index profile, it is necessary to obtain a smoothed envelope of this differential interference fringe. Therefore, as shown in Fig. 8.3-18, the mean values of the fringe shifts and the transverse distances between the two adjacent points were used as the data for the smoothed envelope.

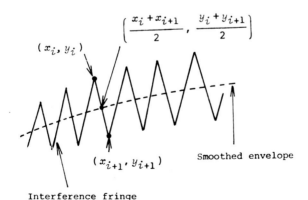

FIGURE 8.3-18.   Smoothing of ripples in the differential interference fringe. (After Y. Kokubun and K. Iga.[56])

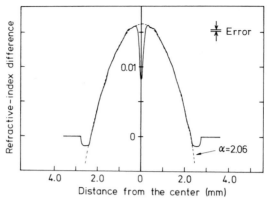

FIGURE 8.3-19.   Refractive-index profile of a rod for a graded-index multimode fiber, where $\lambda = 0.867 \ \mu m$ and $s = 11.3 \ \mu m$. (After Y. Kokubun and K. Iga.[56])

From these data, the refractive-index profile was obtained. The result is shown in Fig. 8.3-19, where the least-squares fit of the exponent α is 2.06.

The refractive-index profile of a fiber drawn from this preform rod was also measured by means of the TDI method. The result is shown in Fig. 8.3-20.

Figure 8.3-21 shows the comparison of the refractive-index profile of the preform rod with that of the fiber. The abscissa is the transverse distance from the center normalized by the core radius. These profiles are almost similar.

Figure 8.3-22 shows the refractive-index profile of the fiber drawn from the same rod, measured by the near-field pattern (NFP) method. In the NFP method, the index valley at the core–cladding boundary cannot be generally observed due to leaky modes. The least-squares fits of α calculated from the

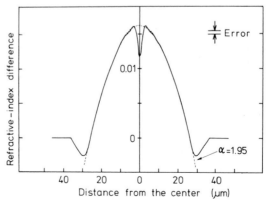

FIGURE 8.3-20.   Refractive-index profile of a fiber drawn from the preformed rod of Fig. 8.3-19 with $\lambda = 0.867 \ \mu m$ and $s = 2.3 \ \mu m$. (After Y. Kokubun and K. Iga.[56])

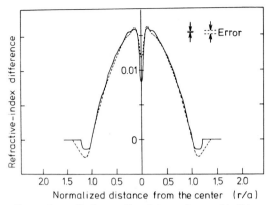

FIGURE 8.3-21. Comparison of index profiles of the preform rod (solid curve) and of the fiber drawn from the preform rod (dashed curve) with $\lambda = 0.867\ \mu m$. (After Y. Kokubun and K. Iga.[56])

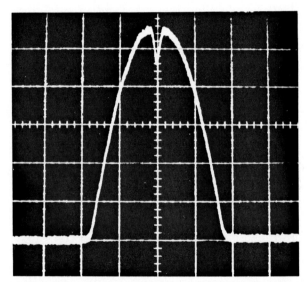

FIGURE 8.3-22. Refractive-index profile of a fiber drawn from the same preform rod, measured by the near-field pattern method. (After Y. Kokubun and K. Iga.[56])

profiles shown in Figs. 8.3-20 and 8.3-22 are 1.95 and 2.05, respectively. These values almost coincide with that of the preform rod. The error resulting from the data reading for profiles measured above was evaluated with the aid of Eq. (8.3-31). The error for the profile shown in Fig. 8.3-19 was ± 0.0002 (1.3% of the index difference between the core and the cladding) and the error for Fig. 8.3-20 was ± 0.0005 (3.3% of the index difference).

I. Refractive-Index Determination of Elliptically Deformed Index Profile

In the previous sections the refractive-index profile was assumed to be axially symmetric. However, the actual fibers and preform rods do not always have an axially symmetric cross section. In that case, the refractive-index profile must be calculated by means of another theory,[77-81] as in the case of tomography in the field of medical electronics. Moreover, if the cross section of a sample has a complicated shape, the three-dimensional index profile must be calculated from the transverse interference or transverse differential interference pattern observed in all directions by rotating the sample. But, if the deformation of the cross section is elliptic, the refractive-index profile can be calculated from the transverse interference pattern observed in only one direction.[82]

In this section the elliptic deformations of some preform rods are measured and the refractive-index profiles calculated from the transverse differential interference patterns. The longitudinal distributions of the elliptic deformation and of the index profile are also discussed.

Figure 8.3-23 shows the transverse differential interference pattern of a preform rod. In contrast with the pattern shown in Fig. 8.3-17, the TDI pattern shown in Fig. 8.3-23 does not include small ripples due to the index fluctuation in each deposited layer except for a ripple corresponding to the center dip. From this it is inferred that the dopants of this preform rod are GeO_2 and P_2O_5. The elliptic deformation of the core was observed by rotating the preform rod sample. The core center was easily determined by observing the sharp ripple in the differential interference pattern due to the center dip of the index profile as shown in Fig. 8.3-23. The core–cladding boundary was determined in the same way. Figure 8.3-24 shows the change of the diameter and the radii of the deposited cladding versus rotation angle. Since the diameter changes sinusoidally, it is seen that this preform rod has an elliptic deformation.

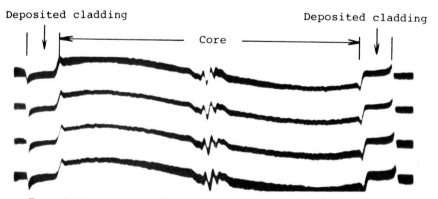

FIGURE 8.3-23.   Transverse differential interference pattern of another preform rod.

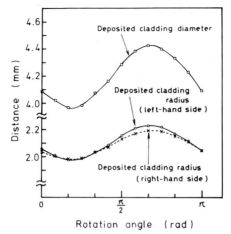

FIGURE 8.3-24. Deposited cladding diameter and radius variation versus the rotation angle. (After Y. Kokubun and K. Iga.[86])

As mentioned above, actual preform rods have some amount of ellipticity. Therefore, it becomes necessary to correct for the effect of elliptic deformation to precisely determine the index profile. Thus far, Chu[82] has proposed a formula which corrects for the effect of elliptic deformation on a transverse interference fringe shift. But this formula includes an integral from zero to infinity at the point of $r = 0$, so it is not suited for numerical calculation.

Now, suppose a fiber or a preform rod sample has an elliptic cross section as shown in Fig. 8.3-25. The refractive index profile is assumed to be represented by $n(r')$. The TI or TDI pattern is assumed to be observed in the direction of either the major or the minor axis. As mentioned in a previous section, the probing ray is actually refracted through the core. In this analysis, however, a

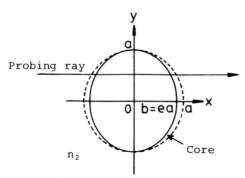

FIGURE 8.3-25. Elliptical deformation of core and ray trajectory, where e is ellipticity, and $r' = \sqrt{(x/e)^2 + y^2}$.

straight-ray trajectory is assumed for simplicity. Then, the transverse inter-ference fringe shift $R(y)$ is given by

$$\frac{\lambda}{D} R(y) = 2 \int_0^{e\sqrt{a^2 - y^2}} [n(r') - n_2] \, dx$$

$$= 2e \int_y^a \frac{[n(r') - n_2]r' \, dr'}{\sqrt{r'^2 - y^2}}. \tag{8.3-33}$$

This is the Abel transform of $n(r') - n_2$, and the refractive index profile is given by

$$n(r') = n_2 - \frac{1}{\pi} \int_{r'}^a \frac{\lambda}{D} \left(\frac{1}{e} \frac{dR(y)}{dy} \right) \frac{dy}{\sqrt{y^2 - r'^2}}. \tag{8.3-34}$$

Since this analysis assumes a straight-ray trajectory, Eq. (8.3-34) includes a maximum error of about 0.8% of the index difference. However, if Eq. (8.3-6), which corrects for the ray refraction, is used to calculate the index profile including an elliptical deformation, the error due to the elliptical deformation is about $1 - e$ of the index difference. Therefore, when the ellipticity is greater than 1%, Equation (8.3-34) should be used to correct for the elliptical deformation.

Figure 8.3-26 shows the refractive-index profiles of the preform rod shown in Fig. 8.3-23 measured in the directions of the major and the minor axes. The number of data samples was about 70. These two profiles coincide. The least-squares fits of α are 1.95 for both profiles.

The most important feature of the TDI method is that the longitudinal variation of the index profile can be measured by this method because it is a

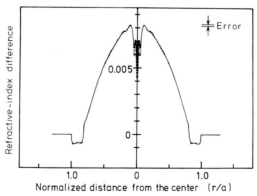

FIGURE 8.3-26. Index profiles of a preform rod measured in the directions of both the major (solid curve) and minor (dashed curve) axes, where $\lambda = 0.867$ μm and $s = 19.3$ μm. (After Y. Kokubun and K. Iga.[86])

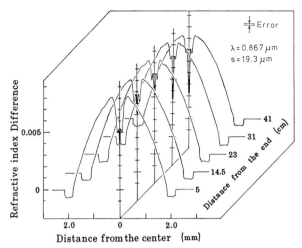

FIGURE 8.3-27. Longitudinal variation of index profile of a preform rod.

nondestructive method. Thus, we measured the longitudinal variation by shifting the sample in longitudinal direction and observing the fringe shift at each shifted point. Figure 8.3-27 shows the measured longitudinal variation of index profile of the preform rod shown in Fig. 8.3-26.

J. INDEX PROFILING OF PLANAR MICROLENSES: SLICED SAMPLE METHOD

The shearing interference method can be applied to sliced samples which have a circularly symmetric index profile such as sliced rod lenses and sliced preform rods. As an example, we sliced the surface of a distributed-index planar microlens[83-85] made of glass into a lateral, thin plate 50 μm thick as shown in Fig. 8.3-28. The sample was illuminated in the direction normal to the surface, and the shearing interference pattern was observed as shown in Fig. 8.3-29 by using the same interference microscope. When it is assumed that the sample is much thinner than the dopant diffusion length and the index profile is radially symmetric, the index profile near the surface of the planar microlens can be obtained from the observed fringe shift as shown in Fig. 8.3-30. The index profile was calculated by solving a difference equation

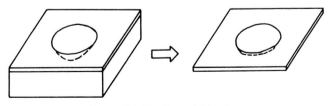

FIGURE 8.3-28. Lateral thin plate.

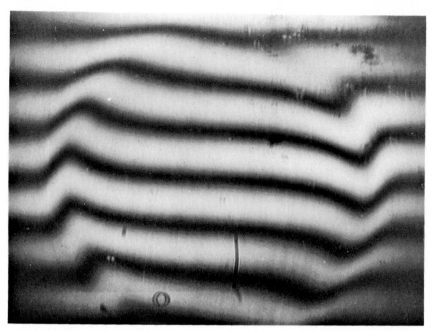

FIGURE 8.3-29. Shearing interference pattern of a sliced sample cut from the surface of a planar microlens. (After Y. Kokubun and K. Iga.[86])

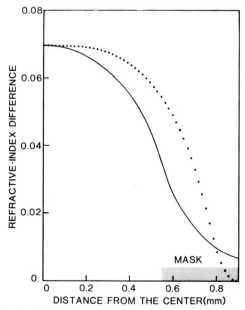

FIGURE 8.3-30. Surface index distribution of a planar microlens, theoretical (————) and experimental (· ·). (After Y. Kokubun and K. Iga.[86])

instead of approximating the difference by the differential of the optical path length. The radius of the mask which was used to prevent the diffusion of dopants is 0.55 mm.[84] The discrepancy between measured and theoretical profiles seems to result from an assumption that the diffusion constant is independent of the concentration of dopants. The diffusion constant may be dependent on the concentration of dopants, or the interaction of the dopants seems to be a possible cause of this discrepancy.

The index profile along the optical axis is obtained by using a sliced sample containing the optical axis perpendicular to the surface as shown in Fig. 8.3-31. In this case, the shearing distance must be larger than the size of the sample. Figure 8.3-32 shows the total shearing interference pattern of this longitudinal thin plate. The thickness was 159 μm. The interference fringes correspond to the cross section of equi index surfaces. Therefore, an index

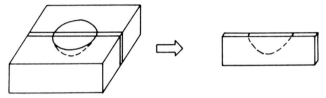

FIGURE 8.3-31.   Longitudinal thin plate.

FIGURE 8.3-32.   Interference pattern of a longitudinal thin plate.

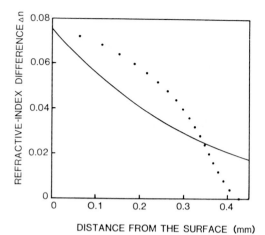

FIGURE 8.3-33. Refractive-index profile of a planar microlens in the depth direction, theoretical (———) and experimental (· · ·). (After Y. Kokubun and K. Iga.[86])

profile in an arbitrary direction can be obtained by scanning in that direction. The index profile in the depth direction was obtained from this pattern as shown in Fig. 8.3-33. The large discrepancy between measured and theoretical curves seems to also be caused by the concentration-dependent diffusion or the interaction of dopants.

Since the planar microlenses measured above were fabricated by the diffusion exchange of dopants, the index distribution was not so hemispherical. Recent planar microlenses were fabricated by means of a deep electromigration technique.[84] Figure 8.3-34 shows the total shearing interference pattern

FIGURE 8.3-34. Interference pattern of a longitudinal thin plate sliced from a planar microlens fabricated by an electro-migration technique.

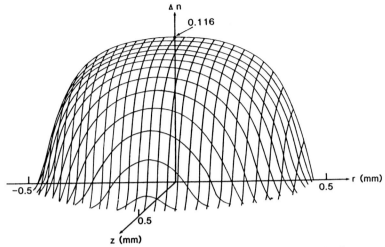

FIGURE 8.3-35. Two-dimensional index distribution of a planar microlens.

of the longitudinal thin plate of a planar microlens that was prepared by a deep electromigration technique. Figure 8.3-35 shows the two-dimensional index profile of this planar microlens. (The planar microlens has a three-dimensional index distribution, but the distribution can be represented by a two-dimensional function of the transverse distance r and the depth z because of symmetry around the optical axis.)

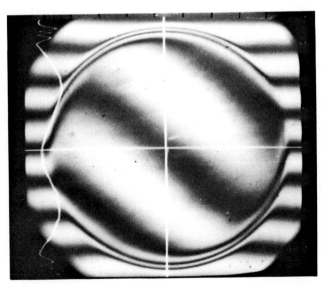

FIGURE 8.3-36. Shearing interference pattern of a planar microlens.

K. Wave Aberration Testing of Planar Microlenses

Since the shearing interference method is based on the measurement of the phase profile at the object plane of an interference microscope, this method can be applied also to the measurement of wave aberration of microlenses.[87] The phase profile at the object plane of the microscope can be reconstructed from the fringe shift. The wave aberration is calculated by comparing the measured phase profile and the reference phase profile of Gaussian focus.

Figure 8.3-36 shows the shearing interference pattern of a planar microlens. The fringe shift is observed by an ITV and is traced by using a microcomputer. The collected data are sent to a main computer. The calculated wave aberration is sent back to the microcomputer and is graphed by an X-Y plotter. Figure 8.3-37 shows the resultant output of the X-Y plotter.

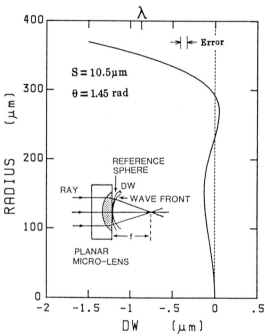

FIGURE 8.3-37.   Wave aberration of a planar microlens.

REFERENCES

1. E. A. J. Marcatilli, *Bell Syst. Tech. J.* **43**, 2887 (1964).
2. S. Kawakami and J. Nishizawa, *J. Appl. Phys.* **38**(12), 4807 (1967).
3. S. Kawakami and J. Nishizawa, *IEEE Trans. MTT* **MTT-16**(10), 814 (1968).
4. T. Uchida, M. Furukawa, I. Kitano, K. Koizumi, and H. Matsumura, *IEEE J. Quant. Electron.* **QE-6**(10), 606 (1970).

5. D. Gloge and E. A. J. Marcatili, *Bell Syst. Tech. J.* **52**(9), 1563 (1973).
6. Y. Suematsu and K. Furuya, *Trans, IECE Jpn.* **54-B**(6), 325 (1971).
7. Y. Suematsu and K. Furuya, *Trans. IECE Jpn.* **57-C**(9), 289 (1974):
8. R. Olshansky and D. B. Keck, *Appl. Opt.* **15**(2), 483 (1976).
9. S. Ishikawa, K. Furuya, and Y. Suematsu, *J. Opt. Soc. Am.* **68**(5), 577 (1978).
10. E. A. J. Marcatili, *Bell Syst. Tech. J.* **56**(1), 49 (1977).
11. A. D. Pearson, W. G. French, and E. G. Rawson, *Appl. Phys. Lett.* **15**(2), 76 (1969).
12. C. A. Burrus, E. L. Chinnock, D. Gloge, W. S. Holden, Tingye Li, R. D. Standley, and D. B. Keck, *Proc. IEEE.* **61**(10), 1498 (1973).
13. W. E. Martin, *Appl. Opt.* **13**(9), 2112 (1974).
14. C. A. Burrus and R. D. Standley, *Appl. Opt.* **13**(10), 2365 (1974).
15. A. H. Cherin, L. G. Cohen, W. S. Holden, C. A. Burrus, and P. Kaiser, *Appl. Opt.* **13**(10), 2359 (1978).
16. B. C. Wonsiewicz, W. G. French, P. D. Lazay, and J. R. Simpson, *Appl. Opt.* **15**(4), 1048 (1976).
17. H. M. Presby, D. Marcuse, and H. W. Astle, *Appl. Opt.* **17**(14), 220 (1978).
18. J. Stone and H. Earl, *Appl. Opt.* **17**(22), 3647 (1978).
19. J. Stone and C. A. Burrus, *Appl. Opt.* **14**(1), 151 (1975).
20. W. Eickhoff and E. Weldel, *Opt. Quant. Electron.* **7**, 10 (1975).
21. M. Ikeda, M. Tateda, and H. Yoshikiyo, *Appl. Opt.* **14**(4), 814 (1975).
22. J. Stone and H. E. Earl, *Opt. Quant. Electron.* **8**(5), 459 (1976).
23. M. Tateda, *Appl. Opt.* **17**(3), 475 (1976).
24. D. Gloge and E. A. J. Marcatili, *Bell Syst. Tech. J.* **52**(9), 1563 (1973).
25. F. M. E. Sladem, D. N. Payne, and M. J. Adams, *Appl. Phys. Lett.* **28**(5), 255 (1976).
26. M. J. Adams, D. N. Payne, and F. M. E. Sladen, *Elec. Lett.* **14**(5), 158 (1978).
27. J. A. Arnaud and R. M. Dersier, *Bell Syst. Tech. J.* **55**(10), 1489 (1976).
28. G. T. Summer, *Opt. Quant. Electron.* **9**(1), 79 (1977).
29. W. J. Stewart, Technical Digest of the International Conference on Integrated Optics and Optical Fiber Communication (IOOC'77), C2-2, (1977).
30. J. P. Hazan, *Electron. Lett.* **14**(5), 158 (1978).
31. K. I. White, *Opt. Quant. Electron.* **11**, 185 (1979).
32. M. Kitsuregawa, H. Nakada, T. Kamiya, and H. Yanai, *TGOQE, IECE Japan* OQE79-65, (1979).
33. H. Kita, I. Kitano, T. Uchida, and M. Furukawa, *J. Am. Ceramic Soc.* **54**(7), 321 (1971).
34. T. Akamatsu, K. Okamura, and M. Tsukamoto, Technical Digest of the Conference on Integrated Optics and Optical Communication, P3 (1977).
35. C. A. Burrus, E. L. Chinnock, I. Gloge, W. S. Holden. Thngye Li, R. D. Standley, and D. B. Keck, *Proc. IEEE.* **61**(10), 1498 (1973).
36. C. A. Burrus and R. D. Standley, *Appl. Opt.* **13**(10). 2365 (1974).
37. S. Hopland, *Electron, Lett.* **14**(24), 757 (1978).
38. A. Ankiewicz, M. J. Adams, D. N. Payne, and F. M. E. Sladen, *Electron. Lett.* **14**(25), 811 (1978).
39. K. Hotate and T. Okoshi, *Appl. Opt.* **18**(25), 811 (1978).
40. T. Okoshi and K. Hotate, *Appl. Opt.* **15**(11), 2756 (1976).
41. E. Brinkmeyer, *Appl. Opt.* **16**(11), 2802 (1977).
42. C. Saekeang and P. L. Chu, *Electron. Lett.* **14**(25), 802 (1978).
43. P. L. Chu, *Electron. Lett.* **13**(24), 736 (1977).
44. K. F. Barrell and C. Pask, *Opt. Commun.* **27**(2), 230 (1978).
45. C. Seakeang and P. L. Chu, *Appl. Opt.* **18**(7), 1110 (1979).
46. M. E. Marhic, P. S. Ho, and M. Epstein, *Appl. Phys. Lett.* **26**(10), 574 (1975).

47. T. Shiraishi, G. Tanaka, S. Suzuki, and S. Kurosaki, *Nat. Conv. Rec. IECE Jpn.* 891 (1975).
48. K. Iga and Y. Kokubun, Technical Digest of the International Conference on Integrated Optics and Optical Fiber Communication, C2-4 (1977).
49. Y. Kokubun and K. Iga, *Trans. IECE Jpn.* **E60**(12), 702, (1977).
50. Y. Kokubun and K. Iga, *Trans. IECE Jpn.* **E61**(3), 184 (1978).
51. C. M. Vest, *Appl. Opt.* **14**(7), 1601 (1975).
52. Y. Maruyama, K. Iwata, and R. Nagata, *Jpn. J. Appl. Phys.* **15**(10), 1921 (1976).
53. K. Nishizawa, M. Tohyama, and T. Fukushige, *Nat. Conv. Rec. Jpn. Soc. Appl. Phys.* **36**, No. 23a-G-1 (1975).
54. K. Iga and N. Yamamoto, *Appl. Opt.* **16**(5), 1305 (1977).
55. Y. Ohtsuka and Y. Shimizu, *Appl. Opt.* **16**(4), 1050 (1977).
56. Y. Kokubun and K. Iga, *Appl. Opt.* **19**(6), 846 (1980).
57. D. Marcuse, *Appl. Opt.* **18**(1), 9 (1979).
58. H. M. Presby and D. Marcuse, *Appl. Opt.* **18**(5), 671 (1979).
59. H. M. Presby, D. Marcuse, and L. G. Cohen, *Appl. Opt.* **18**(19), 3249 (1979).
60. D. Marcuse and H. M. Presby, *Appl. Opt.* **18**(1), 14 (1979).
61. C. Saekeang, P. L. Chu, and T. W. Whitbread, *Appl. Opt.* **19**(12), 2025 (1980).
62. H. Takahashi, S. Shibuya, and T. Kuroha, Technical Digest of the Optical Communication Conference, Amsterdam, 14.4 (1979).
63. I. Sasaki, D. N. Payne and M. J. Adams, *Electron. Lett.* **16**(6), 219 (1980).
64. T. Okoshi and M. Nishimura, *Appl. Opt.* **20**(14), 2407 (1981).
65. J. Winckler, *Rev. Sci. Instrum.* **19**(5), 307 (1948).
66. F. D. Bennett, W. C. Carter, and V. E. Bergdolt, *J. Appl. Phys.* **23**(4), 453 (1952).
67. K. Bockasten, *J. Opt. Soc. Am.* **51**(9), 943 (1961).
68. K. E. Bullen, "An Introduction to the Theory of Seismology," Cambridge University Press, New York, 1963.
69. A. M. Hunter II and P. W. Schreiber, *Appl. Opt.* **14**(3) 634 (1975).
70. G. D. Kahl and D. C. Mylin, *J. Opt. Soc. Am.* **55**(4), 364 (1965).
71. Y. Kokubun and K. Iga, *J. Opt. Soc. Am.* **70**, 36, (1980).
72. K. Hotate and T. Okoshi, *Trans. IECE Jpn.* **E62**(1), 1, (1979).
73. T. Tanaka and Y. Suematsu, *Trans. IECE Jpn.* **E59**(11), 1 (1976).
74. T. Miya, M. Horiguchi, K. Senda, and T. Edahiro, *Nat. Conv. Rec IECE Japan*, 818 (1978).
75. T. Okoshi and M. Nishimura, *TGOOE. IECE Jpn*, OQE79-77 (1979).
76. T. Okoshi and N. Takada, *TGOQE, IECE Jpn*, OQE78-62 (1978).
77. D. W. Sweeney and C. M. Vest, *Appl. Opt.* **11**(1), 205(1972).
78. H. G. Junginger and W. Van Haeringen, *Opt. Commun.* **5**(1), 1, (1972).
79. Y. Maruyama, K. Iwata, and R. Nagata, *Jpn. J. Appl. Phys.* **16**(7), 1171 (1177).
80. P. L. Chu and T. Whitbread, *Appl. Opt*, **18**(7), 1117 (1979).
81. S. Cha and C. M. Vest, *Opt. Lett.* **4**(10), 311 (1979).
82. P. L. Chu, *Electron. Lett.* **15**(12), 357 (1979).
83. M. Oikawa, K. Iga, and T. Sanada, *Jpn. J. Appl. Phys.* **20**, **L51** (1981).
84. K. Iga, M. Oikawa, and T. Sanada, *Electron. Lett.* **17**, 452 (1981).
85. K. Iga, M. Oikawa, and T. Sanada, *in* "Technical Digest of Topical Meeting on Gradient Index Optical Imaging Systems," paper TuB2. Optical Society of America, Hawaii, 1981.
86. Y. Kokubun and K. Iga, *Appl. Opt.* **21**, 1030 (1982).
87. Y. Kokubun, T. Usui, M. Oikawa and K. Iga, *Jpn. J. Appl. Phys.* **23**(1), 101 (1984).

CHAPTER 9

Evaluation and Reduction
of Aberrations
in Distributed-Index Imaging

Studies evaluating and measuring aberrations in rod lenses that have mostly parabolic distributions of their refractive indices are reviewed. Efforts to reduce aberrations by controlling the fabrication process are also summarized.

9.1 OPENING REMARK

A distributed-index (DI) or gradient-index (GRIN) microlens that has a mostly parabolic distribution of its refractive index[1,2] has the potential for introduction into microcomponents used in such optical fiber communications[3,4] and electrooptic systems as copy machines,[5,6] facsimiles,[7] and digital disk systems.[8] In such applications aberration in the distributed-index lenses is of importance[9] because it directly influences the performance of the system. This chapter reviews activities to evaluate and reduce aberrations in distributed-index rod lenses.

9.2 MEASUREMENT AND EVALUATION OF ABERRATIONS

Most of the aberrations in distributed-index lenses are related to the expansion coefficients in the index distribution, which can be expressed, in the case of a clad fiber by [10,11]

$$n^2(r) = n^2(0)[1 - (gr)^2 + h_4(gr)^4 + h_4(gr)^6 + \cdots] \quad (r \leq a)$$
$$= n_2^2 \quad\quad\quad\quad\quad (r \geq a) \quad (9.2\text{-}1)$$

as shown in Fig. 9.2-1a. In the case of a rod lens, there is a scattering wall at $r = a$, as illustrated in Fig. 9.2-1b. Here, $n(0)$ denotes the index at the center axis, g a focusing constant, and $h_4, h_6, \ldots$ are dimensionless parameters.

To summarize some measures which well represent such aberrations, the first group of physical evidence is denoted as aberration coming from the index profile, which is called profile aberration. The second is chromatic aberration.

177

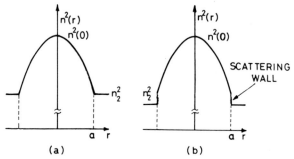

FIGURE 9.2-1   Expression of index distribution. (After K. Iga.[3])

A. DEFORMED GAUSSIAN BEAM

One direct measure of profile aberration is a deformed Gaussian beam transmitted through a distributed-index lens. Suematsu and the author[10] observed a deformed laser beam that was transmitted through a 30-cm-long flow-type gas lens with a nearly parabolic distributed index, as shown in Fig. 9.2-2. In this case the degree of aberration strongly depends on incident beam spot size. Part (a) shows the aberrated beam from a Gaussian beam with

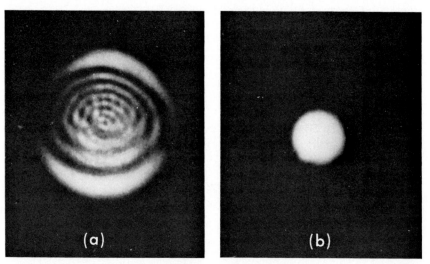

FIGURE 9.2-2   (a) Intensity distribution of a beam astigmatically converted through a flow-type gas lens. The ratio w_{in}/w_0 was much greater than unity. (b) Mode conversion was reduced by making w_{in}/w_0 close to unity. Here w_{in} and w_0 indicate the spotsize of the incident beam and characteristic spotsize, respectively. (After Y. Suematsu and K. Iga.[10])

a large spotsize w_{in}, and (b) shows an output beam that does not have so much aberration because the spotsize is close to the characteristic spotsize w_0, which is equal to $1/\sqrt{k(0)g}$, where $k(0) = n(0)2\pi/\lambda$.

This effect was also pointed out by Kitano et al.[12] for a SELFOC lens and compared with theoretically predicted patterns using the measurement setup shown in Fig. 9.2-3. It is interesting to note that the change in output-beam position is not linearly dependent on the incident-beam position. Also, the shape of the beam deforms with a change in incident position. Maeda and Hamasaki[13] have estimated higher-order coefficients for aberrated laser beams.

B. Pitch of a Sinusoidal Ray Trace

A second measure of aberration is observance of the change in pitch for the sinusoidal trajectory of a meridional ray for which the trace is written as

$$x(z) = x_i \cos \Omega z + (\dot{x}_i/\Omega) \sin \Omega z, \qquad (9.2-2)$$

where x_i and $\dot{x}_i$ are the incident-ray position and slope, respectively. The spatial angular frequency Ω is given by[14,15]

$$\begin{aligned}
\Omega/g = {} & 1 - \tfrac{3}{4}(h_4 - \tfrac{2}{3})\{(gx_i)^2 + x_i^2\} \\
& - \tfrac{3}{4}(h_4 - \tfrac{2}{3})\{21(gx_i)^4 + 46(gx_i)^2 x_i^2 + 17x_i^4\}/12 \\
& - \{\tfrac{3}{4}(h_4 - \tfrac{2}{3})\}^2\{7(gx_i)^4 + 46(gx_i)^2 x_i^2 + 23x_i^4\}/12 \\
& - \tfrac{15}{16}(h_6 + \tfrac{17}{45})\{(gx_i)^2 + x_i^2\}^2.
\end{aligned} \qquad (9.2-3)$$

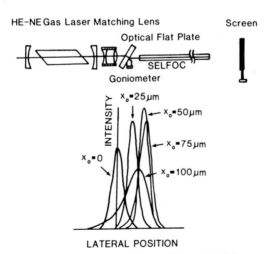

FIGURE 9.2-3   Measurement of the beam position and shape after being transmitted through a Selfoc lens, where x_0 is the position of the incident beam. (After T. Kitano, H. Matsumura, M. Furukawa and I. Kitano.[12])

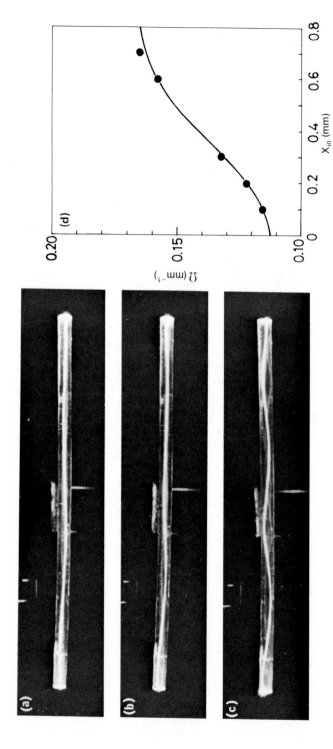

FIGURE 9.2-5 Change of the sinusoidal trajectory of a ray inside a plastic distributed-index rod lens: (a) $x_{in} = 0.1$ mm; (b) 0.4 mm; (c) 0.8 mm; (d) Ω vs. x_{in}, showing measured values (●) and theoretical fit (solid curve). (After K. Iga, K. Yokomori, and T. Sakayori.[16])

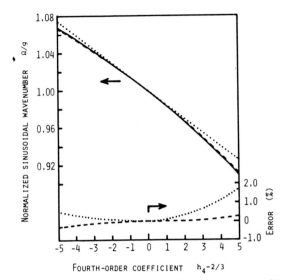

FIGURE 9.2-4   Comparison of normalized sinusoidal wave number Ω/g obtained from Eq. (9.2-3) [dashed curve (sixth order) and dotted curve (fourth order)] with that calculated numerically using the Runge–Kutta method (solid curve) for initial conditions $gx_i = x_i = 0.1$. Dashed curve shows the result obtained by taking only the first and second terms of Eq. (9.2-3) into consideration. (After N. Yamamoto and K. Iga.[15])

Figure 9.2-4 is a comparison of Eq. (9.2-2), with results calculated numerically using the Runge–Kutta method, where $gx_i = 0.1$, $\dot{x}_i = 0.1$, and $h_6 = 0$ were assumed. Related imaging and light-focusing properties can be added to the index profile with the help of Eq. (9.2-2), and this can be utilized to evaluate the DI lens.

Figures 9.2-5a–c show the sinusoidal ray trajectories of a laser beam traveling through a plastic distributed-index rod.[16] The change in Ω against the incident ray position is plotted in Fig. 9.2-5d with h_4 and h_6 estimated to be −203 and 3438, respectively.

C. Imaging Method

The deformed image of the test pattern has been related to the higher-order coefficients by Yamamoto and the author.[15] Figure 9.2-6 shows the measuring apparatus for deformed images of a square-grid test pattern, with higher-order coefficients estimated from experimentally obtained images.[15] This will be detailed at the end of this section.

D. Focusing Method

A focused laser beam is a good measure of the aberrations in a quarter-pitch single lens and is directly related to the performance of the focuser.[8]

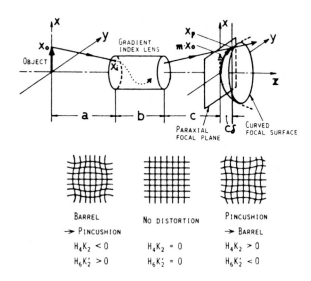

FIGURE 9.2-6.   Imaging method. (After N. Yamamoto and K. Iga.[15])

Figure 9.2-7 shows a focused laser beam from a one-quarter focuser. The theoretical limit of the focused spotsize, which is the diameter of the first zero field, is plotted against the numerical aperture NA in Fig. 5.8-1, where only the fourth-order coefficient h_4 is taken into account.

When focusing a light beam through a parallel plate, the optimum fourth-order coefficient h_4 must be different from the $\frac{2}{3}$ that is the preferable value for a long lens.[11] Discussion that takes h_4 and h_6 into consideration has also been presented by Kikuchi and others.[17] The regions of h_4 and h_6 that provide low lateral aberration have been theoretically obtained.

Ray tracing can immediately show about how big an aberration is. The author presented a computer simulation of a ray tracing where only h_4 was taken into account.[11] If the fourth-order coefficient h_4 is larger than $\frac{2}{3}$, the Gaussian image plane sits in front of the circle of least confusion.

The spot diagram, which is most commonly used in optics, is a nice measure of aberration. The author also demonstrated a one-dimensional spot diagram for changed position of an object.[11] Tomlinson[4] presented amazingly clear spot diagrams generated by a computer. Rawson et al.[18] also reported an aberration expression.

E. Chromatic Aberration

Chromatic aberration is dominant for most imaging applications. Measurement has been made by Ohtsuka et al.[19] for plastic lenses and by Nishizawa[20] for Selfoc lenses, as well as by Gregorka and Moore.[21]

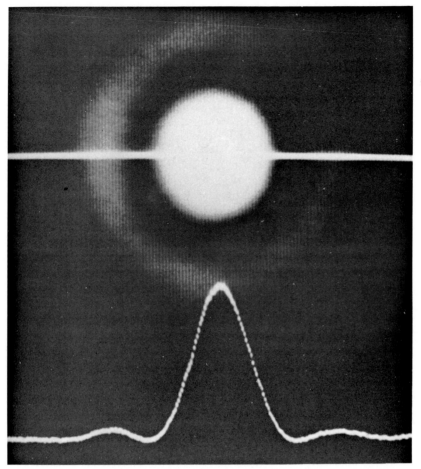

FIGURE 9.2-7.   Focused spot from an experimental SELFOC lens (L_p 14, 2 mm in diameter, $g = 0.304$ mm^{-1}, the $1/e$ spot diameter is 2 μm.) (By courtesy of I. Kitano.)

F. INTERFERENCE METHOD

Direct measurement of index distribution is one effective method of controlling index distribution during fabrication. Chapter 8 summarized some of the methods of measuring optical fibers and distributed-index lenses.

Interference patterns made by a quarter-pitch lens were related to the higher-order coefficients by Rawson et al.[18] A longitudinal interference method applied to 50-μm-thick sliced samples has also been utilized in determining the index distribution of optical fibers and preform rods. This method was introduced by Martin[22] and others. Transverse interference is one method of nondestructive measuring, and a formula that produces an accurate estimation for such has been developed by the authors.[23] Differential interference is quite applicable to a sample that has a large index difference.[24] The accuracy of this method has already been investigated.[25] Reflection,[26] scattering,[27] focusing,[28] and spatial filtering[29] methods are also applicable to identification of index distributions.

9.3 DETAILS OF IMAGING METHOD

A technique is now described to make inspection by imaging fast and easy even during fabrication of DI lens devices. By applying ray tracing, the index profile can be related to the imaging aberrations, and simple expressions of aberrations can be derived analytically. The index profile of a plastic DI lens has been evaluated, and the profile coefficients estimated using this aberration-testing method. Here we consider meridional ray aberration, which is dominant in imaging, compared with skew-ray aberration and wave aberration on the entrance and exit surfaces of the DI lens.

Image formation by a DI lens is illustrated in Fig. 9.2-6. If the DI lens has an ideal refractive-index profile and is thus aberration free, all the rays coming from the object point x_0 are brought to the image point mx_0, where m is the lateral magnification, and the image is formed on a flat paraxial focal plane at distance c from the lens. When the pitch length is reduced to $L_p = 2\pi/g$, the relationship between the image distance c and the object distance a is easily obtained using paraxial approximation.[7] If the index profile is not ideal, however, the image point is still formed most clearly at distance c from the paraxial focal plane, but the clearest image is curved, as shown in Fig. 9.2-6.

A. FOURTH-ORDER COEFFICIENT ABERRATION

If we consider the index distribution in Eq.(9.2-1) to the fourth order, which plays the most important role, the lateral aberration $x_p - mx_0$ and the longitudinal aberration c can be obtained by applying ray tracing in the meridional plane and can be simply expressed by

$$g(x_p - mx_0) = mH_4\{K_1[(gx_i) - \gamma(gx_0)]^3 + K_2(gx_0)^3\}, \qquad (9.3\text{-}1)$$

$$c\delta = cH_4K_3(gx_0)^2, \qquad (9.3\text{-}2)$$

where

$$H_4 = -3(h_4 - \tfrac{2}{3})/4. \qquad (9.3\text{-}3)$$

Aberration coefficients K_1, K_2, and K_3 show the magnitude of three kinds of profile aberrations, corresponding to spherical aberration in a spherical homogeneous lens, distortion, and curvature of field, respectively. Changes in these values as a function of lens length are shown in Fig. 9.3-1. By considering the ray for which the first term of Eq. (9.3-1) vanishes, we can obtain the center position x_p of the sharply focused image point as

$$(\bar{x}_p - mx_0)/mx_0 = H_4 K_2 (gx_0)^2, \tag{9.3-4}$$

which shows that the distortion is proportional to the square of the object height x_0 and to the deviation in the fourth-order coefficient h_4 from the aberration-free value, i.e., $h_4 = \frac{2}{3}$. Therefore, when a mesh with constant line intervals is observed through a DI lens with aberration, the observed image has either barrel or pincushion distortion, as shown in Fig. 9.3-2a. In either case, the focal surface of the sharp image is curved, as shown in Fig. 9.3-2b.

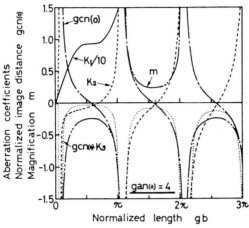

FIGURE 9.3-1　Changes of aberration coefficients K_1, K_2, and K_3, normalized image distance $gcn(0)$, and ideal lateral magnification m against the normalized length gb of a DI lens as a function of the normalized object distance $gan(0) = 4$. (After N. Yamamoto and K. Iga.[15])

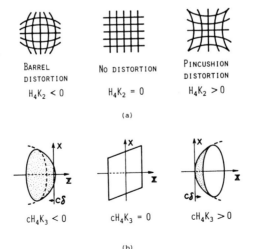

FIGURE 9.3-2　(a) Distortion and (b) curvature of field (to fourth-order coefficient).

B. Sixth-Order Aberration Coefficient

When the fourth-order coefficient h_4 is held close to its optimum value, the sixth-order term in Eq. (9.2-1) becomes dominant on aberrations, which happens most often when using a part far from the center axis for imaging or light focusing. In this case, the right-hand side of Eq. (9.2-4) includes the sixth-order term of x_0, i.e., $H_6 K'_2 (g x_0)^2$, where $H_6 = -15(h_6 + \frac{17}{45})/16$, and K'_2 is the aberration coefficient presenting the amount of distortion of higher order. When both H_4 and H_6 have the same sign, either barrel or pincushion distortion is emphasized. When such a DI lens is used as a light-focusing device, it is difficult to obtain a small spot from a light beam. On the other hand, if they have opposite signs, the observed image includes both types of distortion, as shown in Fig. 9.3-3, and the sharp focal field consists of both convex and concave surfaces.

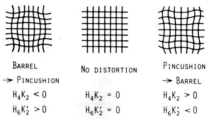

BARREL	No distortion	PINCUSHION
→ PINCUSHION		→ BARREL
$H_4 K_2 < 0$	$H_4 K_2 = 0$	$H_4 K_2 > 0$
$H_6 K'_2 > 0$	$H_6 K'_2 = 0$	$H_6 K'_2 < 0$

FIGURE 9.3-3.   Complex distortion(to sixth-order coefficient) when $H_4 K_2 < 0$ and $H_6 K'_2 > 0$. For example, distortion becomes barrel near-axis and pincushion far off-axis. Aberration coefficient of higher-order K'_2 has the same sign as K_2. (After N. Yamamoto and K. Iga.[15])

TABLE 9.3-I

EXPECTED IMAGE DEFECTS FOR A QUARTER-PITCH LENS

Image		Distortion		Curvature of field
h_4	h_6	Near axis	Far axis	
$> \frac{2}{3}$	$> -\frac{17}{45}$	Pin cushion	Pin cushion	
	$< -\frac{17}{45}$	Pin cushion	Barrel	
$< \frac{2}{3}$	$> -\frac{17}{45}$	Barrel	Pin cushion	
	$< -\frac{17}{45}$	Barrel	Barrel	

Image defects expected from the discussion above are summarized in Table 9.3-I for a quarter-pitch lens. Using the results shown in that table, the index profile can be easily checked merely by observation of an image through the DI lens.

C. Aberration-Testing Systems

A system for evaluation of aberrations is illustrated in Fig. 9.3-4. To measure the sizes of aberrations, a mesh with constant line intervals of 2 mm is employed as the pattern to be observed through the sample lens. The mesh is irradiated by monochromatic light obtained using a monochromator or color filters. By measuring the image distance on the center axis, the focusing constant g from Eq. (9.3-1) can be estimated using the imaging formula for paraxial rays. The amount of distortion was measured by observing each line image formed by the sample lens through a microscope, with a TV monitor used for screening and precise measurement. Index distribution coefficientts h_4 and h_6 were estimated using Eq. (9.3-4) by taking the sixth-order coefficient into account. However, when the index profile approaches the optimum one (aberration free), it becomes difficult to measure the amount of aberration. In that case, a longer sample must be used, since aberrations are magnified every half-period from the image, as can be seen from Fig. 9.3-1. Chromatic aberration can easily be inspected by changing the wavelength of the irradiating light.

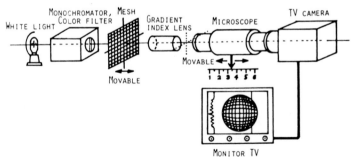

FIGURE 9.3-4 Aberration-testing system used for evaluating aberrations. (After N. Yamamoto and K. Iga.[15])

D. Application to Evaluation of Plastic DI Lenses

As samples for evaluation, we employed plastic DI rod lenses[4,5] fabricated by exchange diffusion of two kinds of plastic monomers with different refractive indices. The plastic monomers used were of lower-index

FIGURE 9.3-5   Mesh image observed by placing a slit 0.3 mm in width in front of a plastic DI lens 8 mm in length and 4 mm in diameter. Distance between the mesh and the lens is 50 mm. (After N. Yamamoto and K. Iga.[15])

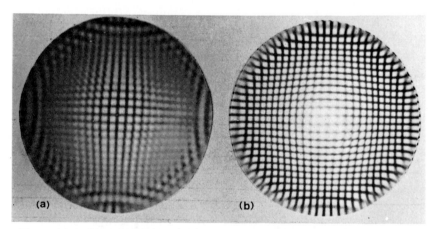

FIGURE 9.3-6   Observed barrel distortion for a quarter-pitch lens with larger aberration. Flat observation field is placed (a) on the center axis and (b) far off-axis. (After N. Yamamoto and K. Iga.[15])

diallyl isophthalate. To investigate the effect of skew rays on imaging, mesh images were also observed by placing a slit in front of the lens, thus eliminating the skew rays. The image observed through the sample lens was 8 mm in length and 4 mm in diameter, as shown in Fig. 9.3-5. The distance from the lens to the mesh was 50 mm, and a slit of 0.3 mm in width was placed 10 mm from the lens. When observation was made without the slit, there seemed to be no change in the amount of distortion or field curvature. Figure 9.3-6 shows the observed barrel distortions obtained by employing a quarter-pitch length sample lens with larger aberration. Since the flat observation field is placed on the center axis in Fig. 9.3-6a, the image near the center is in sharp focus but that far from the axis is quite blurred. Observed pincushion distortion is shown in Fig. 9.3-7. To confirm the reliability of this index profile estimation, we evaluated a sample lens in which the refractive-index profile estimated using transverse differential interferometry[5,6] was $n(0) = 1.547$, $g = 0.192$ mm^{-1}, $h_4 = -0.7$, and $h_6 = 15$. The observed image of a mesh placed 65 mm from the lens is shown in Fig. 9.3-8. The length and diameter of the sample lens were 7 and 2 mm, respectively. Estimated values for the index distribution coefficients obtained from the image shown in Fig. 9.3-8 are $g = 0.192$ mm^{-1}, $h_4 = -1.1$, and $h_6 = 4.4$.

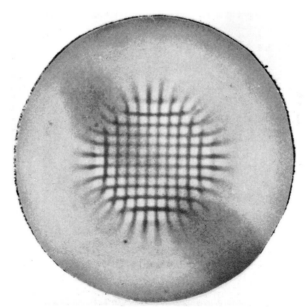

FIGURE 9.3-7 Observed pincushion distortion for a quarter-pitch lens. (After N. Yamamoto and K. Iga.[15])

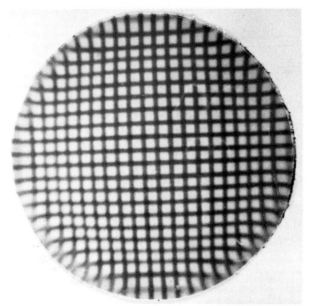

FIGURE 9.3-8.   Observed image of a mesh through a plastic DI lens 7 mm in length and 2 mm in diameter. Object distance between the mesh and the lens is 65 mm, and line interval of the mesh is 2 mm. (After N. Yamamoto and K. Iga.[15])

9.4 REDUCTION OF ABERRATIONS

The index distribution that gives the lowest amount of aberration is obtained by control of the fabrication process, which was already described in Chapter 7.

Figures 7.5-1(a) and (b) show results[30] that present the change of the expansion coefficients h_4 and h_6 of the refractive-index profile against the normalized diffusion time T_3, where D is the diffusion constant and r_0 is the radius of the periphery. In Fig. 7.5-3, longitudinal aberration is plotted against the ion change time measured for experimental SELFOC lenses.[8] The minimum spotsize for a laser beam focused by a thus-controlled lens was 2 μm.

As for methods utilizing diffusion, such as ion-exchange[30] and diffusion polymerization,[16,18] control of not only the diffusion time but also the heat-treatment time is very important[31] in optimizing both h_4 and h_6 at the same time, as shown in Fig. 7.5-2, where the optimum diffusion-exchange and heat-treatment conditions are shown. Fourth and sixth-order coefficients, h_4 and h_6, have optimum values for low aberration within the superimposed region indicated by the arrow, T_2 and T_3 are the diffusion and heat-treatment times, respectively, and D is the diffusion constant and r_0 the radius of the lens.

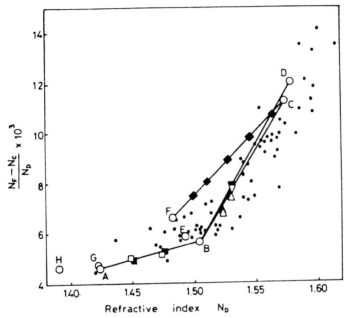

FIGURE 9.4-1 Chromatic aberration and N_D in plastic materials: CR39–3FMA copolymer ($\square$, $\blacksquare$); CR39–DAIP copolymers ($\triangle$, $\blacktriangle$); CR39–VB copolymer ($\triangledown$, $\blacktriangledown$); DAIP–n-BMA copolymer ($\blacklozenge$). (After Y. Ohtsuka, T. Senga, and H. Yasuda.[19])

Chromatic aberrations are based mainly on material dispersion. These are reduced by choosing materials with low dispersion.[18,19] This was demonstrated by Ohtsuka and his colleagues[18] for a plastic distributed-index lens. They measured the Abbe number and refractive index for a number of combinations of plastics, as shown in Fig. 9.4-1. Nishizawa[20] studied the choice of dopants for glass rod lens that give low chromatic aberration. In Fig. 9.4-2 we see the measuring and measured focal lengths of different kinds of lenses. Cesium was found to have low chromatic aberration as a dopant. Gregorka and Moore[21] also demonstrated a Schlieren system for measuring chromatic aberrations in gradient-index lenses, as shown in Fig. 9.4-3.

The combination of focusing and defocusing lens shown in Fig. 9.4-4a may produce achromitization, i.e., compensation for chromatic aberrations.[32] Part (b) represents a designing condition when particular combinations of positive (length L_1) and negative (length L_2) lens pair are assumed.

The theoretical background of aberrations and ways of measuring them are well established. The best lens samples produced in laboratories are approaching a usable level, with high performance, as components in single-mode fiber communication systems and micrometer-spot focusers.

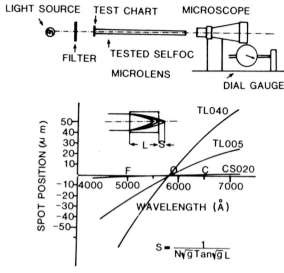

FIGURE 9.4-2   Measurement of chromatic aberration in rod lenses. The microscope measures the spot position *s* which is the distance to the focused spot from the edge surface of the lens. The distance *s* was measured against various wavelengths, and the result indicates the chromatic aberration. (After K. Nishizawa.[20])

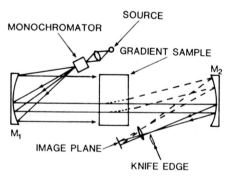

FIGURE 9.4-3   Schematic of the gradient-index schlieren system. The light rays indicated by the solid lines correspond to those without sample. The rays described by the broken lines are for those associated with a sample with gradient index. The change in image position can be measured by changing the wavelength. (After L. Gregorka and D. T. Moore.[21])

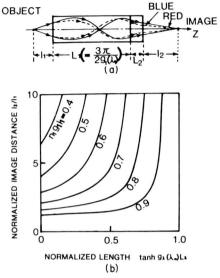

FIGURE 9.4-4. (a) Compensation of chromatic aberration by the use of paired lenses. (b) Achromitization condition. (After K. Iga and N. Yamamoto.[32])

If evaluation and measurement methods that are more suitable for distributed-index lenses are developed and the effort toward reduction of aberration is continued, distributed-index microlenses could find much wider areas of application.

REFERENCES

1. T. Uchida, M. Furukawa, I. Kitano, K. Koizumi, and H. Matsumura, *CLEA*, (1969) *IEEE J. Quant. Electron.* **QE-5**, 331 (1969).
2. A. D. Pearson, W. D. French, and E. G. Rawson, *Appl. Phys. Lett.* **15**, 76 (1969).
3. K. Kobayashi, R. Ishikawa, K. Minemura, and S. Sugimoto, "Fiber and Integrated Optics," Vol. 2, Crane, 1979; T. Uchida and S. Sugimoto, 4th European Conf. Opt. Comm. 8-1 (1978); S. Sugimoto, K. Kobayashi and S. Matsushita, *J. IECE Jpn.* **61**, 1114 (1968).
4. W. J. Tomlinson, *Appl. Opt.* **19**, 1117 (1980).
5. M. Kawazu and Y. Ogura, *Appl. Opt.* **19**, 1105 (1980).
6. K. Matsushita and M. Toyama, *Appl. Opt.* **19**, 1070 (1980).
7. K. Komiya, M. Kanzaki, Y. Hatate, and T. Yamashita, Paper Tech. Group on Visual Communication Engineering, *IECE* Jpn. **IE80-72** (1980).
8. T. Miyazawa, K. Okada, T. Kubo, K. Nishizawa, and K. Iga, *Appl. Opt.* **19**, 1113 (1980).
9. K. Iga, 2nd Top. Meet. on Gradient Index Optical Imaging Systems, Hawaii, 1981; *Appl. Opt.* **21**, 1024 (1982).
10. Y. Suematsu and K. Iga, *Trans. IECE Jpn.* **49**, 1645 (1966).
11. K. Iga, *Appl. Opt.* **19**, 1039 (1980).
12. T. Kitano, H. Matsumura, M. Furukawa, and I. Kitano, *IEEE J. Quant. Electron.* **QE-9**, 967 (1973).

13. K. Maeda and J. Hamasaki, *J. Opt. Soc. Am.* **67**, 1672 (1977).
14. W. Streifer and K. B. Paxton, *Appl. Opt.* **10**, 769.
15. N. Yamamoto and K. Iga, *Appl. Opt.* **19**, 1101 (1980).
16. K. Iga, K. Yokomori, and T. Sakayori, *Appl. Phys. Lett.* **26**, 578 (1975).
17. K. Kikuchi, S. Ishihara, H. Shimizu and J. Shimada, *Appl. Opt.* **19**, 1076 (1980).
18. E. G. Rawson, D. R. Herriot and J. McKenna, *Appl. Opt.* **9**, 753 (1970).
19. Y. Ohtsuka, T. Senga, and H. Yasuda, *Appl. Phys. Lett.* **25**, 659 (1974).
20. K. Nishizawa, *Appl. Opt.* **19**, 1052 (1980).
21. L. Gregorka and D. T. Moore, *Appl. Opt.* **19**, 1096 (1980).
22. W. E. Martin, *Appl. Opt.* **13**, 2212 (1974).
23. Y. Kokubun and K. Iga, *Trans. IECE Jpn.* **E60**, 702 (1977).
24. Y. Kokubun and K. Iga, *Appl. Opt.* **19**, 846 (1980).
25. Y. Kokubun and K. Iga, 2nd Top. Meet. on Gradient Index Optical Imaging System, MD4 (1981).
26. M. Ikeda, M. Tateda, and H. Yoshikiyo, *Appl. Opt.* **14**, 814 (1975).
27. T. Okoshi and K. Hotate, *Appl. Opt.* **15**, 2756 (1976).
28. H. M. Presby and D. Marcuse, *Appl. Opt.* **18**, 671 (1979).
29. I. Sasaki, D. N. Payne, and M. J. Adams, *Electron. Lett.* **16**, 219 (1980).
30. K. Iga and K. Yokomori, *Trans. IECE Jpn.* **58-C**, 283 (1975).
31. K. Iga, N. Yamamoto and Y. Matsuura, *Trans IECE Jpn.* **E60**, 239 (1977).
32. K. Iga and N. Yamamoto, *IECE Japan Nat. Conv. Rec.*, 856 (1977).

CHAPTER 10

Stacked Planar Optics

We propose here a concept involving stacked planar optics, consisting of the stack of arrayed planar microlenses and other optical components. From this configuration a two-dimensional array of optical devices is monolithically available, and many pieces of discrete components can be separated from the array.

10.1 OPENING REMARKS

Great progress in optical fiber communications has been made and many working systems installed.[1] As for the optical components[2] used in optical fiber communications systems, three types have been considered: (i) micro-optics,[3] which consists of microlenses such as gradient-index lenses or tiny spherical lenses, (ii) optical fiber circuits,[4] made from manufactured fibers, and (iii) integrated optics.[5] There have been many problems in the first two schemes, such as optical alignment and fabrication process, and the integrated optics devices are still far from the usable level, as shown in Table 10.1-I.

We have proposed a new concept involving stacked planar optics to overcome these problems. Stacked planar optics consist of planar and two-dimensionally arrayed optical components such as microlenses, filters, and mirrors, and we stack them in tandem to construct functional optical components, such as optical taps, branches, directional couplers, wavelength multiplexers/demultiplexers, and other possible components, including active devices.

One of the key components in stacked planar optics is believed to be a two-dimensional microlens array, but no such array has been formed on a planar substrate. The present authors and collaborators reported the fabrication of planar microlenses[6-8] and have succeeded[9] in making an array of microlenses which might be used in accepting light from optical fibers, i.e., the focal length of each individual 1.0-mm lens 1.6–2.0 mm, and maximum numerical aperture (NA) nearly 0.34.[15] Since the associated acceptance angle is 30°, we believe this claim is reasonable.

TABLE 10.1-I

DESIGN DATA OF PLANAR MICROLENS

	Required	Simple	Pained
NA	0.2–0.5	0.24	0.38
NA_{eff} (aberration free)	0.2	0.18	0.2
Diameter $2a$ (mm)	0.2–1.0	0.86	0.86
Focal distance l (in glass)	0.3–3.8	3.0	1.76
Lens pitch L_p	0.13–1.0	1.0	1.0

10.2 CONCEPT OF STACKED PLANAR OPTICS

Planar optics[10] consists of planar optical components and the stack of them, as shown in Fig. 10.2-1. All components must have the same two-dimensional spatial relationship, which can be achieved from planar technology with the help of photolithographic fabrication as used in electronics. Once we align the optical axis and adhere all of the stacked components, two-dimensionally arrayed components are realized; the mass production of axially aligned discrete components is also possible if we separate individual components as shown in Fig. 10.2-1c. This is the fundamental concept of stacked planar optics, which may be a new type of integrated optics.

A. PROCESS

We therefore propose a possible fabrication process for the stacked planar optics as follows:

(1) design of planar optical devices (determination of thickness, design of mask shape, etc.);

(2) fabrication of planar optical devices;

(3) optical alignment;

(4) adhesion;

(5) connection of optical fibers in the case of arrayed components; and

(6) separation of individual components in the case of discrete components and connection of optical fibers.

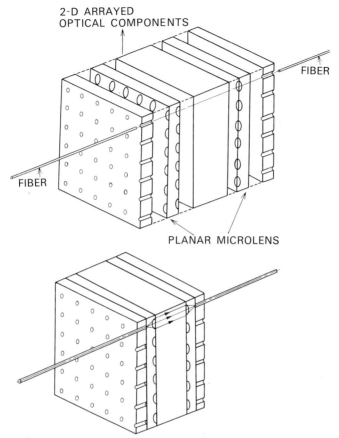

FIGURE 10.2-1. Composition of stacked planar optics. (After K. Iga, M. Oikawa, S. Misawa, J. Banno and Y. Kokubun.[10])

Features of stacked planar optics include:

(1) mass production of standardized optical components or circuits, since the planar devices are fabricated by planar technology.

(2) optical alignment, and

(3) connection in tandem optical components of different materials (glass, semiconductors, electro-optical crystals, etc.) (This had been thought difficult in integrated optics consisting of planar substrates where the connection of different components requires high-precision optical adjustment, since the light is transmitted through a thin waveguide only a few microns in thickness and width.),

(4)   coupling of optical fibers (i.e. it could possibly be done without optical adjustment if precise fabrication of a two-dimensional array of holes is available. This will be detailed later.).

10.3 PLANAR OPTICAL DEVICES

A. Planar Microlenses

To realize stacked planar optics, all optical devices must have a planar structure. The array of microlenses on a planar substrate is required to focus and collimate the light in optical circuits. We have developed a planar microlens[6-8] fabricated by the selective diffusion of a dopant into a planar substrate through a mask, as shown in Fig. 10.3-1. Recently we made an array of planar microlenses with 1.6 to 2.0-mm focal length and numerical aperture (NA) of 0.34. We have confirmed that the NA in the substrate can be extended as high as 0.54 by stacking two of them. This value is considered to be large enough to use it as a focusing light from laser diodes.[16]

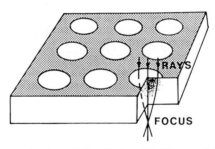

Figure 10.3-1.   Planar microlens. (After K. Iga, M. Oikawa, S. Misawa, J. Banno and Y. Kokubun.[10])

A planar microlens is fabricated by using an electromigration technique, described in Ref. 8. The substrate is a planar glass of $40 \times 40 \times 3$ mm^3, where planar microlenses were formed as an 40×40 matrix with 1-mm pitch. The radius of the mask is about 50 μm and the radius of the resultant lens is 0.45 mm. The improvement of focal length and NA is due to the better choice of mask radius and migration time, and the reduction of a leak current through the edges. The focused spot of the collimated He–Ne laser beam ($\lambda = 0.63$ μm) was measured with the planar microlens. We could observe an Airy-like disk originating from diffraction and aberration, as shown in Fig. 10.3-2. The spot diameter is 3.8 μm, small enough in comparison with the 50-μm core diameter of a multimode fiber, even when we use it in the long wavelength region of 1.3 to 1.6 μm.

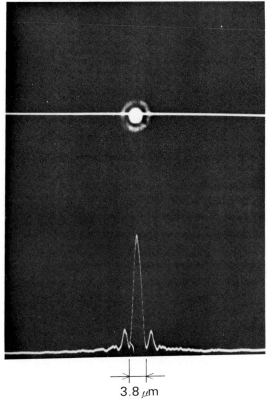

$3.8 \, \mu m$

FIGURE 10.3-2.   Photo of a focused spot by a planar microlens. (After K. Iga, M. Oikawa, S. Misawa, J. Banno and Y. Kokubun.[10])

B. PLANAR OPTICAL DEVICES

In this section we discuss planar optical devices used in stacked planar optics. The actual component configurations are analogous to discrete DI lens systems, but we use these as two-dimensionally arrayed devices which exactly match the lens arrays.

Circular Hole Array   The circular hole array in a planar substrate is used to connect fibers with a stacked planar optical circuit as shown in Fig. 10.3-3a. The diameter of the hole selected is equal to the o.d. of the fiber. Both a penetrating and nonpenetrating hole can be used, as shown in Fig. 10.3-3b and 10.3-3c. The nonpenetrating hole reduces the substrate thickness of a planar microlens. The position of each hole must be matched to that of the arrayed planar microlens to align the optical axis. The alignment process, therefore, can be achieved automatically.

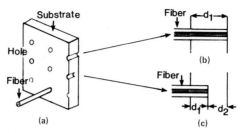

FIGURE 10.3-3.   (a) Array of circular holes showing (b) penetrating and (c) nonpenetrating holes. (After K. Iga, M. Oikawa, S. Misawa, J. Banno and Y. Kokubun.[10])

Aperture and Spatial Frequency Filter   Figure 10.3-4a shows windows opened on a thin film evaporated on a planar substrate. These windows are used as aperture stops. Part (b) shows spatial frequency filters set on the Fourier plane that limit the spatial frequency.

 These devices work to eliminate useless light and to filterout unwanted modes. The lithographic technique can be used to fabricate these components on a substrate or back surface of the planar microlens.

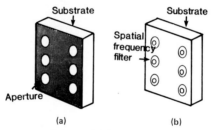

FIGURE 10.3-4.   (a) Aperture and (b) spatial frequency filter array.(After K. Iga, M. Oikawa, S. Misawa, J. Banno and Y. Kokubun.[10])

Wavelength Filter and Mirror   The mirror and wavelength filter on a planar substrate are well known and widely used in conventional optics and optical circuits in optical fiber communication systems.[2] In stacked planar optics, these devices are prepared in a single-batch process with many of the arrayed components. A grating has the same function but must be set aslant,[11] which may require some contrivance to use effectively.

Polarizer and Phase Plate   In a stacked planar optical circuit, a polarizer, analyzer, $\lambda/2$ phase plate, and $\lambda/4$ phase plate can be stacked easily.

Active Optical Components   A component with electrooptic and magneto-optic effects can be applied to deflectors, switches, modulators, and unidirectional waveguides. Although the arrangement of necessary electrodes may sometimes prevent construction of two-dimensional array at the least a one-dimensional array with a stacked configuration can be profitably mass produced. Since the structure of each active device has a special configuration, custom design is called for.

Large NA Lens Array   Since the radiating angle of light from a semiconductor laser diode (LD) and a light-emitting diode (LED) is large (40° for LD, 90° for LED), we need a large numerical aperture lens to accept the light effectively. An arrayed lens with a curved surface, formed from precisely cast plastic or glass and adhered semispheres on a substrate, may be used even though not of planar structure, as shown in Fig. 10.3-5.

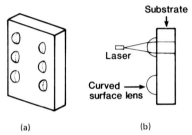

(a)   (b)

FIGURE 10.3-5.   Curved surface lens array. (After K. Iga, M. Oikawa, S. Misawa, J. Banno and Y. Kokubun.[10])

10.4 STACKED PLANAR OPTICS

A. ELEMENTS

Stacked planar optics is analogous to a lens optical system which converts the numerical aperture. In this section we discuss the design of a lens optical system.

Figure 10.4-1 shows three basic optical elements in an optical system. Related matrices introduced in Chapter 6 are also presented. Figures 10.4-1a and b show a buried lens whose thickness is designed equal to the focal length f in a substrate. Basic element A focuses a plane wave on the back surface. On the other hand, basic element B collimates the light from a point source. Basic element C is a homogeneous refractive-index plate with thickness of l. For simplicity, we shall consider the case when the index of each substrate is equal to n.

Since the devices are connected in cascade in stacked planar optics, the ray-matrix method is available to design the system. We use the ray matrix[12] defined by Eq. (10.4-1) with $\dot{x}_1 = \tan \theta_1$ and $\dot{x}_2 = \tan \theta_2$.

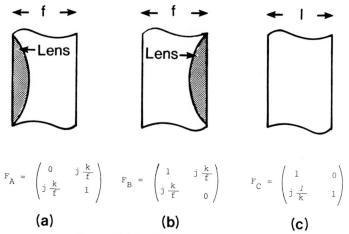

$$F_A = \begin{pmatrix} 0 & j\dfrac{k}{f} \\ j\dfrac{k}{f} & 1 \end{pmatrix} \qquad F_B = \begin{pmatrix} 1 & j\dfrac{k}{f} \\ j\dfrac{k}{f} & 0 \end{pmatrix} \qquad F_C = \begin{pmatrix} 1 & 0 \\ j\dfrac{l}{k} & 1 \end{pmatrix}$$

(a)   **(b)**   **(c)**

FIGURE 10.4-1.   Elements and ray matrices.

$$\begin{pmatrix} jk_1\dot{x}_1 \\ x_1 \end{pmatrix} = (\tilde{F})\begin{pmatrix} jk_2\dot{x}_2 \\ x_2 \end{pmatrix} = \begin{pmatrix} \tilde{A} & \tilde{B} \\ \tilde{C} & \tilde{D} \end{pmatrix}\begin{pmatrix} jk_2\dot{x}_2 \\ x_2 \end{pmatrix}. \qquad (10.4\text{-}1)$$

Here we define x_1 and x_2 as the input and output position, and θ_1 and θ_2 as the angle between the ray and optical axis on the input and output plane in the Nth element as shown in Fig. 10.4-2.

We include the imaginary unit j in the ray gradient to utilize this ray matrix as the matrix which expresses the waveform transformation.[12] The imaginary unit j is eliminated as a common factor in the calculation of ray tracing in the uniform index medium. The inclusion of propagation constants $k_1 = k_0 n_1$ and $k_2 = k_0 n_2$, $(k_0 = 2\pi/\lambda;$ λ is wavelength) expresses Snell's law; that is, if $A = 1$, $B = C = 0$, and $D = 1$,

$$\frac{\dot{x}_2}{\dot{x}_1} = \frac{\tan\theta_2}{\tan\theta_1} \cong \frac{\sin\theta_2}{\sin\theta_1} = \frac{n_1}{n_2}. \qquad (10.4\text{-}2)$$

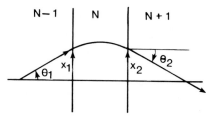

FIGURE 10.4-2.   Transfer of rays ($k_0 = 2\pi/\lambda$). (After Y. Suematsu and H. Fukinuki.[12])

B. Basic Components

A basic component is constructed by combining elements shown in Fig. 10.4-1. We summarize these basic components and their applications in Table 10.4-I.

Coaxial Imaging Component (Table 10.4-I)   The matrix of the coaxial imaging component is obtained by multiplying F_A by F_B:

$$\tilde{F} = \begin{pmatrix} -1 & j\dfrac{k}{f} \\ 0 & -1 \end{pmatrix}. \tag{10.4-3}$$

TABLE 10.4-I

BASIC OPTICAL COMPONENTS FOR STACKED PLANAR OPTICS AND OPTICAL CIRCUITS

Basic components	Application	Reference
Coaxial imaging components	Coupler[a]	4, 11, 12
Noncoaxial imaging components (transmission-type)	Branching circuit[a] Directional coupler[a] Star coupler[b] Wavelength demultiplexer[a]	3, 4, 12, 13 13 12 4, 12, 13
Noncoaxial imaging components (reflection-type)	Wavelength demultiplexor[a] Optical tap[a]	12 12, 13
Collimating components	Branching insertion circuit[a] Optical switch[b] Directional coupler[a] Attenuater[a]	3, 4, 12, 13 12, 13 3, 4, 11, 13 11

[a] Circuit integrated in a two-dimensional array.
[b] Circuit integrated in a one-dimensional array.

By substituting Eq. (10.4-3) into Eq. (10.4-1) we can confirm that the component is the real inverted imaging system, i.e.,

$$x_2 = -x_1, \qquad \dot{x}_2 = -\dot{x}_1 - x_1/f, \qquad (10.4\text{-}4)$$

where we have assumed that the refractive index of the substrate is uniform. In this imaging system, spatial information near $x_1 = 0$ is reconstructed near $x_2 = 0$. When the size of the light source is small ($|x_1/f| \ll \dot{x}_1$), the emergent angle of the ray is nearly equal to the incident angle ($|\dot{x}_2| = |\dot{x}_1|$).

Noncoaxial Imaging System   When the object or light source is not on the optical axis, its image is formed near $x_2 = -\bar{x}_1$. The ray matrix is the same as that of the coaxial system. Since the gradient of the output ray is not only dependent on the incident angle but also incident position (i.e., $\dot{x}_2 = -\dot{x}_1 - \bar{x}_1/f$), the ray proceeds in the direction of $x_1 = 0$ and goes out with the gradient of $-x_1/f$. This different from the terrestrial case, which is realized by a distributed-index rod lens.[11] When the distance between the object and optical axis is small, however, this effect is rather small; for example, if we use a coupled pair of fibers with off-axial value about a half of fiber diameter ($\sim 70\ \mu$m) and a lens of 2.5-mm focal length, the output angle is small ($|x_1/f| \cong 3 \times 10^{-3}$ rad) compared with the angle of light from a fiber ($\frac{1}{6}$ rad).

Noncoaxial Imaging Components (Reflection-Type)   Characteristics of this component are the same as the transmission-type, but the output direction is inverted by a mirror. This component may be used in many optical circuits.

Collimating Components   This component is constructed by inserting a plate between two planar lens elements (Fig. 10.4-1a and b). The total configuration works as an imaging system. The ray matrix before the second lens is obtained by multiplying F_B and F_C, i.e.,

$$\tilde{F} = \begin{pmatrix} 1 - \dfrac{l}{f} & j\dfrac{k}{f} \\[2mm] j\dfrac{l}{f} & 0 \end{pmatrix}. \qquad (10.4\text{-}5)$$

It is found from Eq. (10.4-5) that the incident light is converted into collimated light with $x_2 = 0$ and $x_2 = f\dot{x}_1$. We can insert various planar devices into the collimated region.

In general, we can analyze a stacked planar optical circuit with more layers by using the ray matrices. If we define a matrix of a total system as

$$\tilde{F}_T = \begin{pmatrix} \tilde{A}_T & \tilde{B}_T \\ \tilde{C}_T & \tilde{D}_T \end{pmatrix}, \qquad (10.4\text{-}6)$$

the imaging condition can be obtained from the relation

$$x_2 = mx_1, \tag{10.4-7}$$

where m is magnification, i.e.,

$$\tilde{C}_T = 0 \tag{10.4-8}$$

$$m = \tilde{A}_T, \tag{10.4-9}$$

where $\tilde{A}_T \tilde{D}_T - \tilde{B}_T \tilde{C}_T = 1$. Equations (10.4-8) and (10.4-9) are the necessary conditions of constructing an imaging system.

It might not be necessary to use the matrix method mentioned above in such a simple case shown here. But the matrix is applied to obtain a spotsize and waveform when the propagating light is regarded as a Hermite–Gaussian or Laguerre–Gaussian wave.[12]

10.5 APPLICATIONS

Many kinds of optical circuits can be integrated in the form of stacked planar optics,[14,15,16] as summarized in Table 10.4-I. We do not show optical circuits in detail but refer the reader to Ref. 2.

We present components in this book as examples of stacked planar optics. The optical tap in Fig. 10.5-1 is the component for monitoring a part of light being transmitted through an optical fiber. The problem of optical tap is how to reduce the scattering and diffraction loss at the component. We design an optical tap with stacked planar optics as shown in Fig. 10.5-1. The light from the input fiber is focused by the use of a partially transparent mirror placed at the back surface of the device. Some of the light is monitored with the detector which is prepared on the back of the mirror. The main beam is again focused

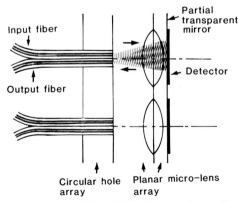

FIGURE 10.5-1. An optical tap with stacked planar optics configuration. (After K. Iga, M. Oikawa, S. Misawa, J. Banno and Y. Kokubun.[10])

by the same lens on the front surface of the output fiber. With this configuration we can fabricate many optical taps on the same substrate. We are actually trying to make this component using fabricated planar microlens. The transmittance of the mirror to the detector was 4.7%. The coupling efficiency of the reflected light was 52%. The position of the fibers and thickness of the substrate is not optimized yet. Coupling efficiency could be improved by solving these problems and reducing the aberrations of the planar microlens.

The 2×3 branching component has been fabricated with two pieces of stacked planar microlenses and half mirror as shown in Fig. 10.5-2.[16] The light from the input fiber was collimated by the first stacked planar microlens, and a part of light was reflected by the half-mirror and focused again to the output fiber 2. The output fiber 2 was set in contact with the input fiber 1. The off-axial value was then 62.5 μm, being the radius of the fiber. The collimated light through the half-mirror was also focussed to the output fiber 3. In order to put the fibers on each surface, the thickness of the planar microlens was carefully designed and adjusted using the ray matrix method as described in Chapter 6. The divided powers were -3.15 and -4.81 dB for output fibers 2 and 3, respectively, when we used a GI multimode fiber of 125-μm outer diameter and 50-μm core diameter. The reflectivity of the half-mirror was 54%, and the transparency was 38% (the absorption loss was 8%). The excess coupling loss was as small as 0.5 dB for output fiber 2 and 0.6 dB for fiber 3.

We were not concerned here with the coupling effect among optical components in the stacked planar optical circuit. But we may construct a

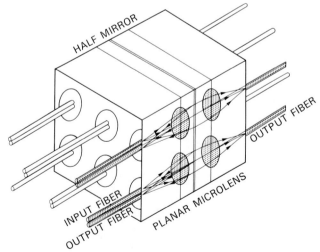

FIGURE 10.5-2. A 2×3 branching component array. (After M. Oikawa, K. Iga, and S. Misawa.)

three-dimensional optical circuit which structures the network by allowing coupling among adjacent components.

Since the accumulation of aberration of lenses may bring about coupling loss, the number of stackings is limited by the aberration of planar microlenses. The reduction of aberration in the planar microlens is important, therefore, if we apply stacked planar optics to more complex components with a large number of stacks.

We have proposed stacked planar optics, a new concept in integrating optical circuits. By using stacked planar optics, we not only make possible the monolithic fabrication of optical circuits such as the directional coupler and wavelength demultiplexer, but we can also construct three-dimensional optical circuits by allowing coupling among individual components in the array with a suitable design.

REFERENCES

1. J. IECE Jpn. Special Issue on Optical Transmission Technology **63**, 1104 (1980).
2. S. Nonaka, J. IECE Jpn. **63**, 1183 (1980).
3. K. Kobayashi, R. Ishikawa, K. Minemura, and S. Sugimoto, Fiber Integr. Opt. **2**, 1, (1979).
4. S. Sugimoto, K. Kobayashi, and S. Matsushita, J. IECE Jpn. **61**, 1114 (1978).
5. Y. Suematsu, J. IECE Jpn. **63**, 1207 (1980).
6. M. Oikawa, K. Iga, and T. Sanada, Jpn. J. Appl. Phys. **20**, L51 (1981).
7. M. Oikawa, K. Iga, T. Sanada, N. Yamamoto, and K. Nishizawa, Jpn. J. Appl. Phys. **20**, L296 (1981).
8. M. Oikawa, K. Iga, and T. Sanada, Electron. Lett. **17**, 452 (1981).
9. M. Oikawa, K. Iga, T. Sanada, S. Misawa, and J. Banno, Fall Meeting of Societies on Applied Physics, 9p-M-2 (Oct. 1981).
10. K. Iga, M. Oikawa, J. Banno, and Y. Kokubun, Appl. Opt. **21**, 3456 (1982).
11. W. J. Tomlinson, Appl. Opt. **19**, 1127 (1980).
12. Y. Suematsu and H. Fukinuki, J. IECE Jpn. **48**, 1684 (1965).
13. S. Sugimoto, R. Ishikawa, and K. Kobayashi, Optics **10**, 128 (1981).
14. S. Misawa, M. Oikawa, and K. Iga, Appl. Opt. **23**, 1784 (1984).
15. M. Oikawa, K. Iga, N. Yamamoto, and T. Yamasaki, ECOC, 107 (1983).
16. M. Oikawa, K. Iga, and S. Misawa, Top. Meet. on Integ. and Guided Wave Opt., ThC5 (1984).

Problems

CHAPTER 1

1. What is the major advantage of the distributed-index lens?
2. Explain the following technical terms in approximately 400 words: lightwave communication; laser disk memory; copy machine using array lenses; and optical sensing. Note how microoptic components are utilized in these fields.
3. What is considered to be an advantage of semiconductor integrated optics?
4. Study the history of distributed index optics. (Refer, for example, to Reference 4 of Chapter 2.)
5. Explain briefly the coupling schemes of waveguides, butt joint, taper coupler, twin-guide or directional coupler, etc.
6. Compare some lightwave components such as microoptic, planar waveguide, manufactured fiber, and stacked planar optical components.

CHAPTER 2

1. Derive Eqs. (2.3-6a)–(2.3-6c).
2. Derive Eqs. (2.3-7a)–(2.3-7c).
3. Explain the meaning of C_i expressed by Eq. (2.3-9).
4. Then try to derive Eqs. (2.3-10a) and (2.3-10b).
5. Derive Eqs. (2.3-15) and (2.3-16). *Hint:* Assume $\partial n^2/\partial\theta = 0$ and first integrate an Euler's equation with respect to z instead of τ in Eq. (2.3-12b) to obtain Eq. (2.3-16). Then substitute Eq. (2.3-16) into the equation for r, similar to Eq. (2.3-12a), and calculate the derivative with respect to z.
6. Derive Eqs. (2.3-22) and (2.3-23).
7. Derive Eq. (2.3-26).
8. Derive Eqs. (2.3-30a) and (2.3-30b). Assume that n^2 is axially symmetric; $\partial n^2/\partial\theta = 0$.
9. Prove Eq. (2.5-6).

10. In approximating Eq. (2.5-6) into Eq. (2.5-4), consider why a can be given by $dx/dz|_{z=0}$ and b by 1.
11. Derive Eq. (2.5-11) by using the perturbation method for solving a nonlinear oscillation equation. (See Reference 12.)
12. Verify Eq. (2.5-16). The refractive index of air is n_a. To obtain n_a outside of the lens use Snell's law.
13. How is n_a affected when the length b of the lens focuser in Fig. 2.5-2 exceeds $0.25L_p$?
14. Obtain Eqs. (2.6-4) and (2.6-5).
15. Obtain the eikonal equation Eq. (2.7-1). (*Hint:* Use Maxwell's equations with $\lambda = 0$.)

CHAPTER 3

1. Consider why a conjugate image can be formed by the array of $\frac{3}{4}L_p$ lenses.
2. The optical transfer function can be defined by

$$H(u) = \int_{-\infty}^{\infty} h(x) \exp(jux)\, dx \qquad (3.P\text{-}1)$$

where $h(x)$ is a function of spot image. Then show that the optical transfer function (OTF) can be expressed by the equation

$$H(u) = \frac{1}{N} \sum_{i=1}^{N} \exp(jux_i), \qquad (3.P\text{-}2)$$

where N is the number of employed rays; and x_i is the position of the ith spot.

3. Consider why the optimum h_4 differs from $\frac{2}{3}$ when we focus light through a transparent plate as shown in Fig. 3.4-3.
4. What is the Wood lens? Also study the Maxwell's fisheye lens.
5. Explain the Gaussian plane and circle of least confusion, which appeared in Fig. 3.4-1.
6. Explain the relation between the fourth-order coefficient of the lens with length shorter than $\frac{1}{4}L_p$ with Gaussian plane and the circle of least confusion.

CHAPTER 4

1. Using Eq. (4.1-2) trace rays incident parallel to the incident plane. (Use a computer.)
2. Derive Eqs. (4.2-10)–(4.2-13).
3. The case in which the modes can be separated into TE and TM modes is limited by the 2-dimensional case; i.e., $\partial/\partial y = 0$. Consider this reason.

4. Obtain the characteristic equation Eq. (4.2-20). Using Eqs. (4.2-20) and (4.2-23) obtain the eigenvalue of b (normalized propagation constant) as shown in Fig. 4.2-2 for TE modes. In solving equations a graphical method may be convenient. *Hint:* Obtain H_z first from Eq. (4.2-1) and then calculate H_z from Eq. (4.2-16).

5. Derive Eq. (4.2-34).

6. Prove that Eq. (4.2.34') stands for both even and odd TE modes when b is known for each mode as a function of V. Show that this is true also for TM modes.

7. Obtain Eq. (4.2-38). Use only the first term in Eq. (4.2-36).

8. To understand the paradox of why the fundamental mode of a step index waveguide has a single peak far-field pattern instead of two peaks, which is expressed by Eq. (4.2-38), obtain Eqs. (4.2-43) and (4.2-44).

9. Explain why the eigenvalue given by Eq. (4.3-12) is limited by the odd integer $2p + 1$.

10. Compare x_0 given in Eq. (4.1-3) with that in Eq. (4.3-7).

11. Show that Eq. (4.3-17) satisfies

$$\int_{-\infty}^{\infty} |E_y|^2 \, dx = 1. \qquad (4.\text{P-}1)$$

12. Obtain Eq. (4.3-18). *Hint:* The Fourier transform of a Hermite–Gaussian function is also Hermite–Gaussian.

13. Refer to a textbook of quantum mechanics to understand Eq. (4.3-20).

14. Verify Eq. (4.3-22).

15. Derive Eqs. (4.4-12) and (4.4-17) from Eqs. (4.4-4)–(4.4-11) *Hint:* Four homogeneous simultaneous equations for A, B, C, and D are obtained from the boundary condition. Equation (4.4-12) is derived from the condition that the determinant of coefficients must be zero.

16. Show that Eqs. (4.4-18) and (4.4-19) express TE and TM modes, respectively.

CHAPTER 5

1. The characteristic spotsize w_0 is given by Eq. (4.5-36) for a round parabolic-index fiber. Using this, obtain Eq. (5.2-3).

2. Obtain Eq. (5.2-5). *Hint:* See Reference 14, 15, or 16.

3. Verify Eq. (5.3-6).

4. Verify Eq. (5.3-10).

5. Show that the Fresnel–Kirchhoff integral in free space can be given by Eq. (5.3-15) by letting g tend to zero.

6. Show that $\phi(x_1, y_1)$ in Eq. (5.4-9) is expressed by 2-D convolution by using Eq. (5.3-14).

7. Verify Eq. (5.4-14).
8. Show that the imaging condition is given by Eq. (5.4-21) and finally by Eq. (5.4-22).
9. Verify Eqs. (5.4-26) and (5.4-29).
10. Prove Eq. (5.7-4).
11. Derive Eq. (5.8-2).
12. Obtain Eq. (5.8-12) and compare to Eq. (5.8-2).

CHAPTER 6

1. Obtain Eq. (6.1-2).
2. Show Eq. (6.1-10) and compare Eq. (6.1-11).
3. Verify Eq. (6.2-6).
4. Obtain the matrices in Table 6.2-1.

CHAPTER 7

1. Derive Eq. (7.2-14b).
2. Show that the solution for drift is given by Eq. (7.3-6).
3. Derive Eq. (7.5-2).
4. Study the formation process of a distributed index by referring to Table 7.7-1.

CHAPTER 8

1. Explain the principle of index measuring methods from Table 8.2-1: longitudinal interference method; near-field pattern method; reflection method; scattering pattern method; transverse interference method; focusing method; and spatial filtering method.
2. Explain the principle of shearing interferometry.
3. What is wave aberration?
4. Show that the solution of Eq. (8.3-5) is given by Eq. (8.3-6).

CHAPTER 9

1. What is chromatic aberration?
2. Explain the achromatization condition in a conventional lens system.
3. What is a major aberration in distributed-index rod lens?
4. Compare the deformed laser beam in Fig. 9.2-3 to the spot diagram shown in Fig. 3.3-2.

CHAPTER 10

1. What is the principle of stacked planar optics?
2. Explain how the lens effect originates in a planar microlens.
3. Explain the lightwave components described in Table 10.4-1: coupler; branch; directional coupler; star coupler; wavelength multiplexer; wavelength demultiplexer; optical tap; and attenuater.
4. List possible applications of stacked planar optics.

Index